WELCOME TO THE GENOME

WELCOME TO THE GENOME

A User's Guide to the Genetic Past, Present, and Future

Rob DeSalle, Ph.D.
Division of Invertebrates
American Museum of Natural History
New York, New York

Michael Yudell, MPH, M.Phil
Molecular Laboratories
American Museum of Natural History
New York, New York

Published in association with the American Museum of Natural History

A JOHN WILEY & SONS, INC., PUBLICATION

For general information on our other products and services please contact our customer Care Department within the U.S. at 877-762-2974, outside the U.S. at 317-572-3993 or fax 317-572-4002.

Wiley also publishes its books in a variety of electronic formats. Some content that appears in print, however, may not be available in electronic format.

Library of Congress Cataloging-in-Publication Data is available.

ISBN: 0-471-45331-5

Printed in the United States of Amercia

10 9 8 7 6 5 4 3 2 1

CONTENTS

FOREWORD

Preventive medicine, individual healthcare, and public health are increasingly dependent on our rapidly advancing knowledge of human molecular genetics and genomics. For many years, we have been aware of several devastating "single gene" diseases like sickle-cell anemia and cystic fibrosis. It has become clear, however, that for most diseases and disorders the genetic causes are rarely this simple. In some cases, genetics play a major role in either the vulnerability to develop a specific disease, or in the body's protection against disease development. It is also increasingly evident that gene-environment reactions are of seminal importance. Scientists now recognize that our genetic composition does not represent predestination or predetermination, but rather sets the stage for the interaction between genes and an array of factors in our individual and shared environments. This relationship can influence our risks of developing any disease or disorder. In another area of importance, our world food supply is increasingly dependent on our ability to manipulate genes of plants and animals of all types, desirable in many cases, but controversial in others.

The new science of genomics is having a growing impact on our lives. However, the public's knowledge about genes, molecular genetics, and human genomics is extraordinarily limited, both in the United States and throughout the world. This knowledge gap is not surprising, both because of the technical nature of genomics and since many of the critical developments in this field have been made only recently and, thus, after the years of any formal education of the majority of adults. Even for those who may have had the privilege of learning about the fundamental aspects of molecular biology, molecular genetics, or genomics in their formal high school or college education, it is difficult to keep up with the enormous volume of new information and its applications and implications. Therefore, a book such as this one, *Welcome to the Genome: A User's Guide to the Genetic Past, Present, and Future*, provides a welcome primer for individuals of diverse ages, educational backgrounds, and current interests.

The story told here, that is, the story of molecular genetics and genomics, is well over a century old. In 1869, Friedrich Miescher discovered and elucidated the chemistry of nucleic acids, which was a critical first step in this story. However, it was not until the extraordinary work in the early nineteen-forties of Oswald Avery, Maclyn McCarty, and Colin MacLeod (briefly detailed in this book), working at The Rockefeller University, that there was the first proof that nucleic acids are the centerpiece of the transmission of "traits between generations" and, thus, that nucleic acids are the "fundamental elements of life." (Dr. Maclyn McCarty recently celebrated his ninety-third birthday and is vigorous and actively participating in the intellectual ferment of The Rockefeller University community, many of whose members are working in either molecular biology or molecular genetics, including studies related to the human genome and its global implications.) In 1953, the pivotal studies of James Watson and Francis Crick, along with colleagues in England and in the United States, unraveled the structure and the function of DNA. Since that time, all biological science has been radically changed and is undergoing rapid evolution, a process which will undoubtedly continue for many years to come.

In the human genome, we now appreciate that all of humankind living today is extraordinarily similar. On an individual level our large genomes differ by only 0.1%. Or, in other words, we are 99.9% genomically alike. However, variants of specific genes which play important roles in our biology are neither "good" nor "bad," but simply variations in the human genome. Although these differences do not reflect the ways in which we have historically divided humankind, they are proving scientifically and medically significant. For instance, studies of the human mu opioid receptor gene, where many of our "endorphins" bind, as well as where pain medications (such as morphine) and drugs of abuse (such as heroin) act, have allowed us to identify a genetic variant which plays an enormous role in many different aspects of human physiology. The physiological changes that may result from one copy of a variant can be measured objectively, are substantial in magnitude, and may have implications for development of specific diseases. Despite the "sameness" of human kind, there is obvious diversity. Increasingly, we appreciate that this diversity may have implications, not only for diseases, but also for response to medications and possibly for understanding our own physiology.

Many readers may have concerns about the ethical issues surrounding genomic science. These concerns may increase as the reader learns of the many areas where genetics and genomics are beginning to play a role in our lives. At the personal level, and with respect to each of our own genetic information, many or most of us would like to have knowledge of our own genome if it can make us healthier. At the same time, most of us would also want to be assured that no negative consequences result from that knowledge. It is absolutely essential that no "reprisals," or unenlightened negative actions, be taken based on genetic information, such as the presence of genetic variants that may contribute to cancer, cardiovascular disease, depression, or an addictive disease. It is essential that there be national and international legislation and, hopefully, international standards and ethical constraints by which no denial of insurance or inappropriate escalation of insurance fees, nor denial of access to education, employment, or advancement therein be based on genetic composition. As you will learn in this book, a majority of U.S. states offer varying protections against genetic discrimination, and the U.S. Senate in 2003 passed bipartisan legislation making genetic discrimination in health insurance and employment illegal. Each individual adult needs to understand the scientific basis, including clinical implications, of genetic information. Each individual also has to contemplate, on the one hand, the personal "desire to know", as well as the "need to know" by physicians and other healthcare providers and, in that context, the question of who has the "right to know." Similarly, debates are ongoing nationally and internationally concerning the use of genetically modified foods. It is important to have these debates. However, they need to be based on a fundamental base of information and knowledge. Again, this book offers an overview and highlights elements of the ethical and other controversies that have been engendered by rapid developments in the field. *Welcome to the Genome*, provides a wonderful resource of information. While I am sure this will be one of many books for the lay public on this topic to be published in future years, DeSalle and Yudell offer a uniquely informative perspective on the genomic revolution.

Mary Jeanne Kreek, MD

Patrick E. and Beatrice M. Haggerty Professor

Head of Laboratory

The Laboratory of the Biology of Addictive Diseases

The Rockefeller University

Senior Physician

The Rockefeller University Hospital

ACKNOWLEDGMENTS

All books are collaborations of some sort. Vital to the completion of a manuscript are an editor's encouragement and skill, colleagues' good will, friends' tolerance for reading chapters, and loved ones' patience and nurturing. Authors are also indebted to their sources and to the many librarians and archivists who guide them during their search for information. For us, as isolating as the writing process could sometimes be, it was ultimately a team effort.

We are thankful to the many members of our team who made this book possible. At the American Museum of Natural History Maron Waxman, formerly Special Publications Director, who is now enjoying retirement, has been a dedicated editor, staunch advocate, and good friend for several years now. Without her and her assistant Curtis Brand, this book would not have been possible. The staff of the Museum's photo studio, including Denis Finnin and Craig Chesek, helped us put together many of the images seen throughout the text. Some of these images began as components of the Museum's exhibition "The Genomic Revolution." We are grateful to the Museum's Exhibition team, including Vice President for Exhibitions David Harvey and exhibition designer Tim Nessen.

Perhaps the greatest gratitude we owe for their inspiration and contributions to the book belongs to the writers and researchers in the Exhibition Department. Lauri Halderman, Karen de Seve, and Martin Schwabacher were the writers for "The Genomic Revolution." Working closely with Rob, the exhibition's curator, and with Yael Wyner, the content coordinator, Lauri, Karen, and Martin developed a language to talk about the genome to Museum visitors in a way that made complex science both accessible and interesting. Their work had a significant impact on our approach to and on portions of this book. Some of the passages and case studies in many of the following chapters were originally researched and written by the exhibition's writing team. We especially relied on their approach and writing in chapter 8, which looks at genetically modified organisms. And we must also thank and recognize Yael Wyner for her efforts in guiding the exhibition's content and thus for her intellectual contributions to this book.

The entire staff of the Museum's library and archives deserves our special thanks for helping us with this project. Finally, we want to thank Museum President Ellen Futter and Museum Provost Michael Novacek, whose continued support and commitment to public education about the genome project helped get this book off the ground.

Luna Han, our editor at John Wiley & Sons, has been ceaseless in her dedication to this book, and we are continuously grateful to her. From her initial interest in this project more than two years ago, Luna has given us her constant support as she expertly cultivated our manuscript from beginning to end. Editors generally play various roles in a book project, and Luna filled all of these admirably, at once nurturing the text, being a constructive critic, and forcing us to meet or come close to meeting sometimes-trampled deadlines. Danielle Lacourciere, Associate Managing Editor at Wiley, worked closely with us to complete the production of this book. We remain appreciative of her hard work.

Our photo editor, Karin Fittante, spent scores of hours tracking down many of the images you see in this book. We greatly value her perseverance and patience in contacting archives and individuals to find photos and secure photo rights.

Three colleagues read the manuscript in its entirety. Dr. Dennis Liu, Program Director at the Howard Hughes Medical Institute, gave us page-by-page comments on the content of the book. His insight into the subject matter often spurred us to think more deeply about genomics and revise our text accordingly. We owe him our deepest gratitude. We also owe important thanks to Karen Miller, Director of Editorial Services at the American Museum of Natural History, who lent her critical eye and red pen to the effort. Karen's editorial skills greatly improved the quality of this book. Finally, Dr. Brian Reed, postdoctoral fellow in the Laboratory of the Biology of Addictive Diseases at the Rockefeller University, provided critical insight into many of the technical issues in the book. We offer him our deepest appreciation.

Many colleagues and friends also read one or many chapters of the book or spoke with us at length about issues in the text, and we thank them for their valuable comments. They are Dr. David Rosner, Michael Begleiter, Dr. Gerald Oppenheimer, Dr. Samuel Wilson, Dr. Richard Sharp, Elizabeth Ribolotti, Dr. James Bonacum, Dr. Mary Egan, Dr. Patrick O'Grady, Julian Stark, Dr. Michael Russello, Dr. Mary Jeanne Kreek, Dr. Howard Rosenbaum, Adrienne Burke, Neil Schwartz, Bill Shein, Sandy Kandel, Andrea Yudell, Bette Begleiter, Paul Messing, Alice Pifer, Pamela Lyss, Avinoam Patt, Larry Rudman, Deborah Sacks, and Kelvin Sealey. Also, Rosalie Goldberg, a genetic counselor, provided invaluable assistance early on in the research and writing of this book. We also benefited from discussions with Evan Wilcox about genetically modified foods and gained a deeper understanding of some of the criminological uses of DNA from Aliza Kaplan at the Innocence Project.

We would also like to thank for their continued support all of the members of the DeSalle Molecular Systematics Laboratory at the American Museum of Natural

History and the faculty and students in the Center for the History and Ethics of Public Health at Columbia University.

We are also extremely fortunate to have had one another during this process. A cooperative project can have its drawbacks—two very different schedules and two sets of commitments sometimes slowed us down. But we found ways to complement our strengths in the production of this book, and together we completed our task.

For two summers we provided technical content and instruction at the summer genomics institute for high school teachers sponsored by the Woodrow Wilson National Fellowship Foundation. Parts of many of the chapters in this book were first developed as lectures for the institute and we are grateful for the feedback we received from the teachers there. We are thankful to both Nancy Anderson, who ran the institute our first year, and Rob Baird, who ran it our second year.

Finally, our families and loved ones deserve special thanks. They have, after all, sustained us during this process with their unyielding love and affection. Rob thanks his wife Annie Williams and his daughters Miriam and Sonya. Michael thanks his parents Allen and Jane Yudell and his sister Andrea Yudell for their continued support and love. Michael also thanks Jacqueline Rick, whose passion for life, intellectual enthusiasm, and adoring love made the last few years during the often rigorous writing of this book both manageable and remarkable.

INTRODUCTION:
WELCOME TO THE GENOME

Every one of the trillions of cells in your body contains DNA—from the blood cells that course through your veins to the nerve cells in your brain to the hair follicle cells that line your scalp. The tightly coiled DNA in a single cell, six feet long and just one molecule wide, packs more than three billion bits of information. This complete set of information is your genome. Approximately 30,000 genes in your genome program the development of a new human being and are constantly at work instructing our bodies to create new cells, digest food, fend off disease, and store thoughts. Genes and DNA capture our imagination because of their impact on why we are the way we are. But how much control do genes and DNA really have? And to what extent will our increasing understanding of the human genome change who we are and how we see the world? Are our genes our destiny? Are our genomes our fate?

Such questions capture our imagination in the midst of the genomic revolution—the international multi-billion-dollar effort to sequence, interpret, and exploit the human genetic code. A map of our genome offers boundless potential to scientists. Foremost are prospects for our health, ranging from identifying disease susceptibilities to discovering cancer cures. We can also apply genomic information to feeding the world's growing population, solving forensic mysteries, and saving species on the verge of extinction. Almost daily the media report new genetic discoveries from the frontier of biology. Headlines like "Long Held Beliefs Are Challenged by New Human Genome Analysis," "Genome Shows Evolution Has an Eye for Hyperbole," and "Double Helix Is Starting To Make Its Mark In Medicine" underscore the genome's complexity, allure, and promise. (1) Despite the current popularity of genetics, it is daunting to grasp the idea that a gene, which is a simple and elegant molecular amalgamation of nucleic acids, the "stuff" of biological heredity, has so much control and influence over our lives. The nature of that control and its scientific, medical, and social meanings are the subjects of intense debate taking place in laboratories, in corporate boardrooms, at cocktail parties, in the halls of Congress, and in houses of worship.

The meanings and mechanisms of heredity were pondered and debated millennia before the development of modern genetics. In the fifth century BCE, the Greek dramatist Euripides wrestled with the complexities of the relationship between parent and child in his play *Electra*:

> I oft have seen,
> One of no worth a noble father shame,
> And from vile parents worthy children spring,
> Meanness oft groveling in the rich man's mind,
> And oft exalted spirits in the poor. (2)

Without knowledge of genes or genomes, premodern thinkers had many ideas concerning the nature of heredity, some of which were surprisingly sophisticated and accurate. To Euripides heredity must have been a mystifying and seemingly random process. How else could he and his contemporaries explain the inconsistencies among inherited traits within families? Other ancients carefully considered similar questions. Lucretius, a Roman philosopher, wrote that traits could skip generations, as children sometimes resembled their grandparents. (3) Around the globe, premodern farmers had already developed sophisticated breeding techniques that depended, in part, on a basic understanding of heredity. We know, for example, that the ancient Assyrians and Babylonians artificially pollinated date palm trees and that many animals, including sheep, camels and horses were domesticated during ancient times. (4) The domestication and breeding of plants and animals shows that many early thinkers recognized that traits were passed between generations.

Perhaps the most advanced premodern thinker on heredity was Aristotle (384–322 BCE). (5) Aristotle dedicated much of his work to questions concerning the specific mechanisms of heredity. He theorized that inherited traits were passed between generations by what he called the *eidos*, or the blueprint, that gave form to a developing organism. Aristotle's *eidos* was entirely theoretical—he could not see this invisible configuration—a fact that makes his theory all the more remarkable. Aristotle understood the mechanisms of heredity only in the broadest sense and remained handicapped by the limited technology of his time, a primitive understanding of biology, and the cultural limitations of his worldview. Yet a keen perception, buttressed by his emphasis on observation and description, made him a brilliant interpreter of the natural world. The concept of the *eidos* remained the most complete theory of heredity until the modern era of genetics. More than two millennia later scientists use a genetic language strikingly similar to Aristotle's. The *eidos* is in many ways analogous to the modern concept of a genome, and like Aristotle today's scientists often refer to a genome as a blueprint for life. (6)

Craig Venter (left), President Bill Clinton, and Francis Collins (right) at the White House announcement on June 26, 2000 of the completion of the draft sequence of the human genome. Venter and Collins fiercely competed against one another to be the first to finish the sequence. An eleventh-hour truce was arranged, and they decided to call their race for the human genome a tie.

It has been a journey to today's understanding of heredity. Genetic knowledge and the exploration of heredity from Aristotle to the sequencing of the genome, built over the centuries, are yielding information that can be applied to scientific theory and medical practice. During the twentieth century, science translated a philosophy and theory of inheritance into the molecular biology that made possible the sequencing of humanity's DNA. The Human Genome Project and other biotechnology initiatives are bringing about radical changes in medicine, agriculture, and the study of our evolutionary heritage as well as creating new social, ethical, and political dilemmas. In June 2000, scientists triumphantly announced they had sequenced the human genome, the blueprint for human life. (7) By sequencing the 3.2 billion units of our DNA, researchers sparked a firestorm of discovery and ushered in a new age.

At a White House ceremony to announce the completion of a draft sequence of the human genome, President Bill Clinton called the genome God's handiwork. "Today," Clinton stated, "we are learning the language in which God created life." (8) Clinton's vision of the genome was one that mixed a metaphor of scientific advancement with a divine spirit. This image of the human genetic code is a fairly common one. The genome has also been called the book of life, biology's Rosetta stone, humanity's instruction manual, and biology's Holy Grail. Each of these metaphors

In all, the human genome contains over 3 billion units of DNA code, arranged in a fixed sequence that defines the human species. These three stacks of 142 bulky phone books are filled only with Gs, As, Ts, and Cs—the letters in the your genetic code. Believe it or not, all of this genetic information is packed into the nucleus of every cell in your body.

conveys a slightly different meaning, and each suggests a subtly different aspect of the genome. Not so hidden in these metaphors is the hope that biology will provide clear-cut answers to long-asked questions regarding the nature of the human soul, the power of science to heal and rebuild the human body, and the role of nature in human social behavior. The genome will indeed provide some answers to these questions, but not the simple answers that many of these metaphors suggest.

Genomics is a synthesis of many disparate fields, including biology, public health, engineering, computer science, and mathematics. What makes genomics even more distinctive is that the social sciences and humanities are an integral component of the genomic revolution. Philosophers, ethicists, and historians are helping to lay the foundation of the genomic revolution by pushing for and playing a role in the creation of policies and laws that will guide the integration of genomics into scientific practice and health care. Participants in the genomic revolution, as well as the biologists and others who preceded them, will, we believe, be thought of much in the same way that Newton is remembered for his role in the birth of calculus and physics or the way in which Darwin is remembered as the progenitor of modern biology. However, because genomics is an evolving science that encompasses so many different disciplines, it is hard to find one person who embodies the entire field. Indeed, it will be a group of genomic scientists who will be recorded in history books as pioneers.

The arrival of the genomic age was the culmination of efforts of over a century of science. From the work of Gregor Mendel in the mid-nineteenth century (it was Mendel who formalized the rules of heredity and hypothesized that something like genes must underpin heredity) to the announcement of the discovery of the structure of DNA in 1953 by James Watson and Francis Crick to the genetic sequencing technologies developed by biologists like Frederick Sanger and Leroy Hood in the closing decades of the twentieth century, the path to genomics has been arduous but has yielded the richest source of biological data we have ever known. This age of discovery is where our journey in this book begins—the first few chapters look at the historical moments in biology over the past 100 or so years that made the sequencing of genomes possible. These chapters will be particularly rewarding to readers with an interest in the science behind genomics, but you do not need to comprehend everything in these chapters to appreciate the material in the rest of the book. Don't get hung up on some of the nitty-gritty science. Utilize the figures to help make sense of difficult concepts, and liberally use the glossary at the back of the book. And, finally, look forward to getting to the book's remaining chapters. They are not as scientifically challenging.

The last two parts of the book will look at the interplay of how scientists are coming to make sense of genomic information and how they are applying this information to genomic technologies in health care and agriculture. Part II, entitled "Information," will look at how the discovery and exploration of the human genome is yielding to the more practical task of sorting through the scientific and social meaning of all of the data being generated by genomics. The choices, social proscriptions, and laws that we develop now around genomic technologies will be an essential part of ensuring the success of genomic technologies in the future. Challenges include creating policies that will help integrate genomics technologies into contemporary medicine and public health practice, and defining the roles and responsibilities of scientists, health care professionals, ethicists, clergy, and lawmakers in the development of these policies. Also, how can we best ensure the safety of genomic technologies? Part III, "Advancement," will look at how genomics is becoming a part of our lives. For example, diagnostic technologies are already available to test for a variety of genetic conditions, and the first molecular targeted drugs such as Gleevec, a treatment for a form of leukemia, are just now coming to market. (9) But it will still take years, possibly decades, before genomic medicine will significantly enhance current practice, let alone replace it.

We have set out to write a book that readers with little or no prior knowledge of biology can pick up and enjoy, gaining along the way a deeper understanding of the phenomenon that has become known as genomics. Genomics should not be treated lightly, however, and we hope to reward your interest with more than a nominal exploration of this burgeoning science. Indeed, one can pick up any number of magazine or newspaper articles for that. This book offers something more—something useful to you, the consumer, by elucidating today's genomic information and tomorrow's genomic medicines and technologies. It is the latter that will, in various ways, greatly affect our lives. Although we may not directly benefit from incredible genetic discoveries, children born in this new century will come into the world with the promise that genomics will have a significant impact on their lives, and for their children the effect will be exponentially greater, continuing likewise through the generations.

For us, though, the consequences of genomics will be no less significant. Although we will benefit from early generations of genome-driven therapeutics, we also face the critical task of struggling with the consequences of these potentially disruptive technologies. We are charged with making sense of the genome's social, cultural, and economic implications and with successfully implementing genome technologies. Although lives will be improved and even saved by genomic drugs, our generation's legacy will be much more than the scientific and medical discoveries it leaves to the twenty-first century. Our legacy will also be social—meeting the challenge of making genomics technologically feasible and at the same time humane, just, and ethical. This will be no easy task, particularly from our current vantage point: At present we as a society remain largely unprepared for the arrival of the genomic revolution. This book was written with these challenges in mind, and with the hope that we can be a part of the continued effort to make the genome truly public.

At the American Museum of Natural History we have been working toward integrating genomics into Museum scientific practice and into our exhibits. In the fall of 2000, as part of our ongoing mission to bring cutting-edge science to the public, the Museum held a two-day conference examining the social and scientific implications of the genome. *Sequencing the Human Genome: New Frontiers in Science and Technology* was the first major public forum to examine the implications of genomics after the release of the draft sequence of the human genome. Renowned scientists, including two Nobel laureates, bioethicists, historians, biotechnology entrepreneurs, and others participated in a variety of lectures and panel discussions. This effort was followed in spring 2001 with the opening of the exhibition *The Genomic Revolution,* the largest and most comprehensive popular examination of the genome to date. Efforts continue through the Museum's education programs and by expanding the reach of *The Genomic Revolution,* which is traveling to several sites around the United States.

For well over a century the Museum's halls, replete with fossils, models, and dioramas, have been home to a diversity of exhibitions that, with few exceptions, have centered on objects—exactly the fossils and dioramas that fill the Museum's galleries. These object-driven exhibits utilize the charisma of a specimen to engage the visitor. An ancient *Barosaurus* standing on its hind legs, towering forty feet in the air does just that in the main rotunda of the Museum every day. Once a visual connection to a specimen is made, the conceptual aspects of an exhibit can be presented. In the case of the *Barosaurus,* the Museum can discuss a wide range of such dinosaur-related topics as predation, evolution, and extinction. The specimen draws in the visitor, but precisely because of that charismatic attraction he or she leaves with a much deeper understanding of dinosaurs.

The Genomic Revolution approached the art of exhibition-making and museum education in a much different fashion. Instead of relying on the allure of an object, the genomic revolution itself, in its abstract and complicated splendor, is what attracted the visitor. The physical specimens were secondary to theories, ideas, and scientific premises. The challenge for the exhibition team lay in translating these difficult concepts into dynamic and decipherable objects that illustrate the genome. To meet this task a team of Museum scientists, experts in the field, and exhibition specialists grappled with the problems for well over a year before delivering *The Genomic Revolution.* The striking success of the exhibition suggested to us that charisma is not necessarily object based, and for our purposes here, that was encouraging.

For this book, a dinosaur example is again useful. Looking at a pteradon skeleton hanging by wires from the Museum ceiling, its wings outstretched in flight, our imagination takes us to a prehistoric era when dinosaurs ruled. But for the genome our imaginations are used in a much different way. Genes are, in essence, invisible to us. Imagining molecular processes may be of use to a geneticist or biochemist,

The forty-foot *Barosaurus* welcomes visitors every day to the American Museum of Natural History in New York City. This amazing specimen immediately draws visitors into the lives of dinosaurs.

but for the rest of us picturing the activities of nucleic acids, DNA, and genes is a challenging, if not futile, exercise. The charisma of the genome lies instead in its possibilities, not simply in what a molecule of DNA can do, but in what DNA can do for us—its potential to change humanity and our environment in ways once only dreamed of. Therein lies the public's fascination with the genome and with other biotechnologies.

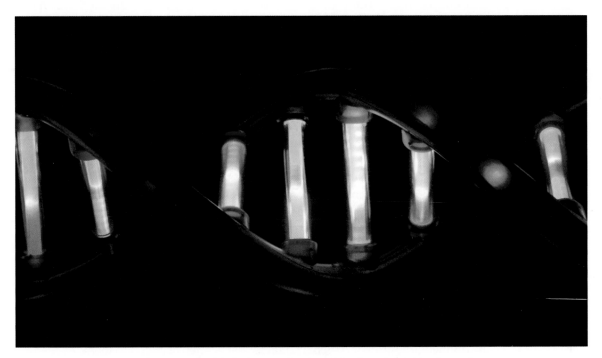

This artist's conception of a DNA double helix was displayed in the exhibit *The Genomic Revolution.*

This book, which is neither a synopsis of the exhibition nor an accompanying text, hopes to capture this charisma. To help accomplish this task we utilize imagery from the exhibition to illustrate many of our points and to provoke your molecular imagination to think about the amazing things that our bodies and the molecules of which we are composed can do. The richness of many of the photographs you will see in this book is testament to the tremendous efforts of the Museum's exhibition team, to whom we owe great thanks for making our task that much easier.

Despite popular and sometimes scientific opinion to the contrary, genes are not the determinative force that many contend or hope they are. Claims of genetic control over intelligence, sexuality, and aggression have come and gone and will come and go again. However, although genes unquestionably contribute to behavioral and medical outcomes, they generally do not govern how we behave or entirely control what diseases we contract or develop. There is a tendency to confuse genetic destiny and genetic potential—a confusion that lies in our changing understanding of gene function. For nearly a century the dominant paradigm in human heredity theory boasted that traits were inherited via single genes (or loci). Scientific support for a one-gene, one-trait approach in genetics was, in fact, borne out by many of the genetic discoveries of the twentieth century. It was easy to show, for example, that certain traits are directly inherited through the mechanism of a single gene. Devastating diseases such as sickle-cell anemia, Huntington disease, and Tay–Sachs disease could all be pinpointed to a single locus. Ultimately, this approach has been fruitful only in the simplest cases of inheritance. The inheritance of these types of diseases is rare, probably accounting for "no more than 5% of known disease."(10) Yet, this single-gene, single-trait approach still holds sway in some scientific circles and remains ubiquitous among the general public, despite science's failure to genetically understand common and stubborn diseases such as cancer, heart disease, and diabetes, all of which claim many lives each year, and all of which have complex etiologies that are both genetic and environmental. If genetics in the twentieth century was about the search for origins of human traits gene by gene, then twenty-first century genomics is about the transition away from single-gene thinking. Genomic technologies are opening up new ways of thinking about the mechanisms of heredity.

Genes are not destiny, and such an assertion undermines the astonishing complexity and possibility that are our genes. But if the role of genes in our lives is not this simple, then why read any further? After all, you are reading a "user's guide" to your genes and may have been expecting us to tout the wonders of our genetic code. We are enchanted by the genome and its potential to change our lives in so many ways, but there is so much more to genes and the genomic revolution than the divinelike control and global panacea that is often ascribed to them. By reading this book you will learn about the myths and realities of the genome, and in doing so prepare yourself to be an educated participant is the incredible changes to come. In the coming pages you will learn about, among other things, pharmacogenomics and personalized medicine, genetic engineering and agricultural biotechnology, and the role of genomics in helping us better understand our evolutionary heritage. We must remember that the sequence of the human genome is only a first step, and that despite the promises ahead, genomics is still in its infancy. It is likely that we cannot even envision some of what is to come, our imaginations lacking the technological and biological prowess to see a future beyond science fiction. Educating ourselves about the genome will no doubt improve our visionary skills and empower us to be participants in these amazing times. Putting the genome to work raises questions and dilemmas for us as individuals, families, nations and even as a species. We need to make decisions about our health, our food, our stewardship of the natural world, and our responsibilities to the next generation.

Welcome to the genome.

PART 1

Discovery

For much of the nineteenth and all of the twentieth century, scientists worked to unravel the biological basis of inheritance. With Gregor Mendel's mid-nineteenth century discovery of the basic mechanisms of heredity, genetics was born, and humanity took its first small steps toward deciphering the genetic code. No longer would heredity solely be the domain of philosophers and farmers. Indeed, Mendel's discoveries set the stage for major advances in genetics in the twentieth century and help put in motion the series of discoveries that have led to the sequencing of the human genome. This age of discovery, from Mendel to the sequencing of the human genome, is the subject of this part of the book. Chapter 1 covers some basic biology and tells the story of the evolution of genetics by examining some of the most significant discoveries in the field—discoveries that enabled the development of genomics. Chapter 2 looks specifically at the development of genetic and genomic sequencing technologies. Finally, Chapter 3 examines the human genome itself and the ways in which we are exploring and exploiting it now and in the future.

From Mendel to Molecules

Without any further ado, may we present to you the human genome!

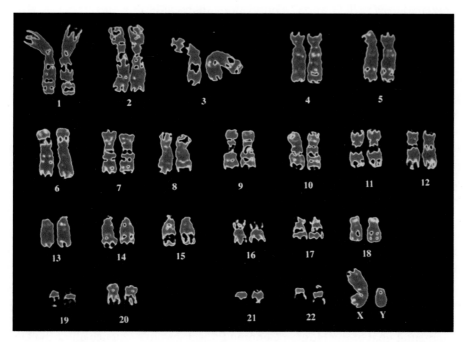

This picture, known as a karyotype, is a photograph of all 46 human chromosomes. With an X and a Y chromosome, this is a male's karyotype. A female's karyotype would show two X chromosomes.

This photo, also known as a karyotype, shows the 46 human chromosomes, the physical structures in the nuclei of your cells that carry almost the entire complement of your genetic material, also known as your genome. In almost all the cells in the human body there are 22 pairs of chromosomes and two sex-determining

3

chromosomes. The double helices that make up your chromosomes are composed of deoxyribonucleic acid, also known as DNA, on which are found approximately 30,000 genes. These cells are called somatic cells, and they are found in almost all nonreproductive tissue.

Humans also have cells with 23 nonpaired chromosomes. In these cells each chromosome is made up of a single double helix of DNA that contains approximately 30,000 genes. These cells are called germ cells and are the sperm and egg cells produced for reproduction. These germ cells carry a single genome's worth of DNA or more than three billion base pairs worth of nucleic acids.

Chromosomes are somewhat like genetic scaffolding—they hold in place the long, linearly arranged sequences of the nucleotides or base pairs which make up our genetic code. There are four different nucleotides that make up this code—adenine, thymine, guanine, and cytosine. These four nucleotides are commonly abbreviated to as A, T, G, and C. Found along that scaffolding are our genes, which are made from DNA, the most basic building block of life. Through the Human Genome Project scientists are not simply learning the order of this DNA sequence, but are also beginning to locate and study the genes that lie on our chromosomes. But not all DNA contains genes. The long stretches of DNA between genes are known as intergenic or noncoding regions. And even within genes some DNA seems to be nonfunctional or "junk" DNA. These areas, called introns, are interspersed within the functional parts of a gene, known as exons. The term junk DNA may turn out to be a misnomer. Some scientists hypothesize that these noncoding regions of DNA may play a role in regulating gene function. (1) Unlike the human genome and all other eukaryotic genomes, bacterial genomes do not have introns and have very short intergenic regions.

Let's begin our tour of the human genome with a very basic lesson in genetic terminology. For example, what exactly is genetics, and how is it different from genomics? Genetics is the study of the mechanisms of heredity. The distinction between genetics and genomics is one of scale. Geneticists may study single or multiple human traits. In genomics, an organism's entire collection of genes, or at least many of them, are examined to see how entire networks of genes influence various traits. A genome is the entire set of an organism's genetic material. The fundamental goal of the Human Genome Project was to sequence all of the DNA in the human genome. Sequencing a genome simply means deciphering the linear arrangement of the DNA that makes up the genome.

In animals, the vast majority of the genetic material is found in the cell's nucleus. The Human Genome Project has been primarily interested in the more than 3 billion base pairs of nuclear DNA.

A tiny amount of DNA is also found in the mitochondria, a cellular structure responsible for the production of energy within a cell. Whereas the human nuclear genome contains more than three billion base pairs of DNA and at least 30,000

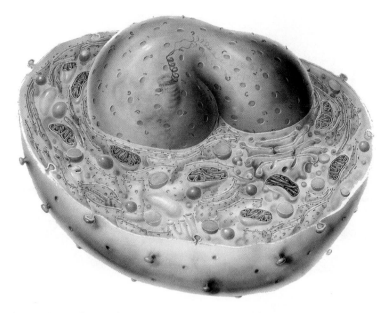

The nucleus of every human cell (the large purple mass inside the cell)
contains DNA. Mitochondria, organelles in cells that produce energy
(the smaller purple objects within the cell), also contain some DNA.

genes, the human mitochondrial genome contains only 16,568 bases and 37 genes.
(2) Like bacteria, mitochondrial DNA, or mtDNA, has short intergenic regions and
its genes do not contain introns. Another interesting characteristic of mtDNA is
that it is always maternally inherited. This has made mtDNA very helpful to track
female human evolutionary phenomena. These discoveries were made possible, in
part, by sequencing mtDNA.

What about heredity? In the most basic sense we should think about heredity as the transmission of traits from one generation to the next. When we talk
about heredity in this book we refer to the ways in which traits are passed between
generations via genes. The term heredity is also sometimes used to describe the
transmission of cultural traits. Such traits are shared through a variety of means including laws, parental guidance, and social institutions. Unlike genetics, however,
there are no physical laws governing the nature of this type of transmission. Only
genes can carry genetic information between generations.

What are genes? Genes are regions of DNA and are the basic units of inheritance in all living organisms. These words, genes and DNA, are too often used
interchangeably. Both genes and DNA are components of heredity, but we identify
genes by examining regions of DNA. In other words, DNA is the basic molecular
ingredient of life whereas genes are discrete components of that molecular brew.

Although identical twins (top) share the same genome, there are always physical and behavioral differences between them. Fraternal twins (bottom) may look alike, but they do not share the same genome. Nontwin siblings also do not share the same genome.

If you look at any family you'll see both shared and unique traits. Family members typically look alike, sharing many features such as eye color and nose shape, but they may also have very different body types and be susceptible to different diseases. This diversity is possible for two reasons. The first reason is that genes come in multiple forms. These alternative forms are known as alleles, and in sexual reproduction they are the staple of organismal diversity. According to the laws of genetics siblings can inherit different traits from the same biological parents because there is an assortment of alleles that can be randomly passed along. The second reason is that the environment can exert a significant influence on the expression of genes. For example, an individual may inherit a gene that makes him or her susceptible to lung cancer. Such susceptibility is typically revealed, however, only after years of genetic damage caused by cigarette smoking or other lung-related environmental impacts. (3)

So how *did* science progress from thinking about the mechanisms of heredity to understanding that genes are the basic units of heredity, to deciphering and finally manipulating the DNA code that underlie all life on Earth? The results of the Human Genome Project were the fruits of over a century of struggle by scientists around the globe. Most historians of science would measure this progress beginning with Gregor Mendel's work on pea plants during the middle of the nineteenth century. Although premodern thinkers did have a basic grasp of the idea of heredity—that is, that identifiable traits could be passed down from generation to generation—it was not until Mendel that science began to understand the mechanisms underlying the transmission of these traits. (4)

The journey from abstract notions of inheritance to the sequencing of the human genome abounds with stories of discoveries both great and small that led to where we are today. Science seldom progresses in a straight line. The genome was always there for us to find but took centuries to discover because knowledge and the technological application of that knowledge advance fitfully, revealing gradually more over time. Scientists have not always made the right choices. Even today, in the midst of the genomic revolution, we may be making assumptions about our genes that future generations look back on and ask, "How could they have thought that?" The trials and errors of science are part of what makes this process so interesting.

Several major building blocks of life had to be discovered to make possible our entry into the genomic world. First, scientists needed to determine what constitutes the hereditary material that passes from one generation to the next. Second, they needed to find out what constitutes the biochemical basis for the expression of this intergenerational legacy. This endeavor required the ability to take cells apart and analyze the chemical components from different parts of cells. Scientists then needed to determine the ways in which these chemicals, the building blocks of life, interacted, how they were structured, and how that structure influenced the hereditary process. Finally, technologies needed to be developed to use this information to improve human health, agriculture, and our understanding of our place in the history of life on Earth.

It took almost 150 years from the discovery of the hereditary principles to the sequencing of the human genome. The stories behind these discoveries explain how scientists came to understand the biological basis of heredity. What follows does not represent the comprehensive history of all the important genetic work of the last century or so. Yet without the discoveries we highlight, here the discovery of the genome would never have occurred or would have happened very differently.

IN THE ABBEY GARDEN

For close to two millennia few scientists approached Aristotle's understanding of genetics. Other theories of heredity were put forth during the centuries. Some, like the idea of the homunculus—the belief that every being was miniaturized and preformed in a reproductive cell—or the belief in panspermia—the idea that secretions from the entire body contribute to offspring—held sway for great lengths of time. (5) But it was not until the Austrian monk Gregor Mendel bred peas in his abbey garden that anyone made practical sense of the rules of heredity.

Mendel was not just a monk tending peas. The child of peasant farmers, he was a classically trained scientist raised in the greatest traditions of the Enlightenment. Intellectually nurtured by his family and schooled in the best academies and universities of Central Europe, the German-speaking Mendel spent his life dividing his affection between God and science. (6) In 1843, at the age of 21, Mendel entered the St. Thomas Monastery in Brünn in what is now the Czech Republic. (7)

In the Church Mendel found a community of scientists—botanists, zoologists, and geologists among them—working diligently in their fields and making important contributions to the scientific literature. Perhaps the most important event in Mendel's early career occurred 10 years into his stay at St. Thomas. In 1851, at the behest of his abbot, Mendel was sent to Vienna University to study at the institute of Professor Christian Doppler, one of the pioneers of modern physics. For two years at Doppler's institute Mendel honed his scientific skills, taking courses in physics, chemistry, and mathematics, as well as entomology, botany, and plant physiology. The influence of physics was important to Mendel's later work on heredity. Physics taught Mendel that laws governed the natural world and that these laws could be uncovered through experimentation. (8) But it was ultimately Mendel's exposure to ongoing debates in heredity that transformed him into the scientist we remember today.

Mendel and his predecessors understood that traits could be passed between generations. A child with his mother's eyes and his father's nose was easy evidence of that. Breeding experiments with domesticated animals also suggested that traits

Although it took decades for Gregor Mendel's work on pea plants to revolutionize hereditary theory, his impact is today still felt in the biological sciences.

were passed to offspring. The prevailing theory during the nineteenth century, one to which even Charles Darwin mistakenly ascribed, was "blended inheritance." (9) This theory held that the characteristics of parents blended in their offspring. Experimentation in this area failed because, as Mendel was able to eventually determine, heredity was not a lump sum but rather a series of individual traits.

In 1856 Mendel began to study the mechanisms of inheritance, working with varieties of garden peas from the genus *Pisum*. (10) In the course of his experiments his garden flowered, as did his understanding of heredity. Mendel discovered several generalities from his experiments that remain the foundation of twentieth-century genetics. Any student of biology knows Mendel's work. Known as Mendel's laws, these basic tenets describe heredity in two simple mechanisms: the law of independent assortment and the law of segregation.

Mendel began an experiment with purebred peas. One breed had yellow seeds, the other green seeds. When purebred yellow-seeded peas were bred with each other, their offspring through the generations would have yellow seeds. Under the same circumstances, the green-seeded peas would always have green-seeded progeny. However, when he bred the purebred pea with yellow seeds to a pure-bred pea with green seeds, the offspring, or the first generation of this breeding cross, always had yellow seeds. The green seed trait seemed to be gone. Mendel called traits like the yellow seed trait dominating (now called dominant) because in first-generation crosses they would always appear. (11) Traits like the green seed trait were called recessive—although they disappeared completely in the first generation, they reappeared in the second. Thus when Mendel took the yellow seeds from the first generation and either self-pollinated them or pollinated them with pollen from other yellow peas from the same first-generation breed, he discovered that some of these offspring, the second generation, again had the green seed trait. The plants, Mendel concluded, retained the ability to produce green seeds—of the second-generation seeds, 6022 were yellow and 2001 were green. Likewise, when he used six other traits, he found the same pattern in the second generation—traits that had disappeared in the first generation reappeared in the second. (12) The chart below shows the relationship between dominant and recessive traits in second-generation pea plants in the seven traits Mendel experimented with.

Dominent Trait		Recessive Trait	
Round seeds	5474	Wrinkled seeds	1850
Yellow seeds	6022	Green seeds	2001
Gray seed coats	705	White seed coats	224
Green pods	428	Yellow pods	152
Inflated pods	882	Constricted pods	299
Long stems	787	Short stems	277
Axial flowers	651	Terminal flowers	207

(13)

From these experimental data Mendel made several conclusions that are at the heart of his revolutionary contribution to hereditary theory. From the 3 to 1—dominant to recessive—ratio in the second generation, Mendel concluded that the traits he studied came in two different forms and that these forms existed in pairs in the plant. Mendel called these forms factors. Today we call them genes. During the process of making reproductive cells, Mendel deduced, these genes segregate from each other—that is, the two copies of a gene that you get from each parent segregates, and in the subsequent reproductive cells, only one half of the pair is passed on to offspring. At fertilization, a gene from each parent reconstitutes the pair. How else could Mendel explain how two yellow-seeded pea plants could produce offspring with green seeds? In this case the green seed trait was as much a part of the pea plant as the yellow seed trait despite sometimes being hidden. Mendel also concluded that the factors that were dominant (in the left-hand column above) somehow overcame the factors that were recessive (in the right-hand column) when they were combined in offspring from crosses. When all first-generation plants were crossed, they had both kinds of factors. Mendel's calculations allowed him to predict a 3 to 1 ratio if the two factors were segregated. This is Mendel's first law, the law of segregation, which states that the factors specifying different alleles are separate or segregated, that only one may be carried by a gamete (an egg or sperm), and that gametes combine randomly. Therefore, a child has the same chance of inheriting allele A as it does allele B. (14)

Without the assistance of a calculator or computer, Mendel counted thousands and thousands of plants. Even more remarkable, he constructed lineages that had all possible combinations of two of the seven traits together. For example, he crossed a line of pea plants with round yellow seeds with a line whose seeds were wrinkled and green. This cross gave rise to first-generation plants with seeds that were all yellow and smooth. But when he crossed these first-generation plants to each other (a self-cross), an amazingly regular ratio in the offspring arose—the seeds of nine were yellow and round, three yellow and wrinkled, three green and round, and one green and wrinkled. Mendel reasoned that mating these first-generation plants was like taking the two possible types of each trait (for example, seed texture and seed color) and throwing them into a hat. Nature then randomly chose from the hat how to combine the genes. Although the choice is random, the outcome is

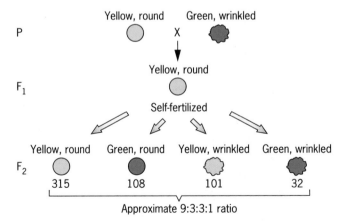

Mendel's first law, the law of segregation, says that alleles will segregate *randomly* between generations. Mendel's second law, the law of independent assortment, represented in the figure above, says that pairs of alleles will segregate *independently* between generations. (P = Parents, F_1 = First Generation, F_2 = Second Generation)

a remarkably regular ratio—9:3:3:1. (15) These observations are now known as Mendel's second law, the law of independent assortment—if two traits (genes) are being controlled with different controllers (alleles), offspring will be produced by random combinations of the controllers (alleles). (16) In other words, a trait is independently and randomly distributed among offspring.

Mendel was either very lucky or very perceptive; it turns out that seven is the number of chromosomes of *Pisum*. For all seven of the traits he examined to show true independent assortment with respect to one another, none of them can be linked—that is, none of them can be on the same chromosome (or in the case of one of the traits he examined, they have to be very far apart on the same chromosome). (17) Mendel must have watched his peas very closely. Perhaps he recognized the pattern of segregation as he was weeding his garden and thus performed his experiment with an expectation based on his knowledge as a pea biologist. Or, perhaps, he selectively looked at his data and forgot to record crosses that deviated from the ratios 3:1 and 9:3:3:1. Either way, his conclusions have not been overturned.

Mendel died on January 6, 1884, nearly 20 years after his momentous study with *Pisum* had been published. (18) Even though its significance remained unheralded, Mendel's work as a scientist and as a servant of God was recognized by his peers. If Mendel had been luckier in choosing the journal in which to publish his findings, he might have been famous in his own time, but he chose an obscure scientific journal and died in genetic obscurity (his monastic calling guaranteed that). (19) Although his contemporaries did cite his work with *Pisum*, they probably did not comprehend its deeper meanings for what would become a cornerstone of hereditary theory. A tribute to Mendel by a fellow scientist in Brünn lauded him as one of the great scientists of his day who dedicated himself "almost exclusively on detailed natural scientific studies, in which he displayed a totally independent, unique way of thinking." (20) Unfortunately, it would take the world another 16 years after his death to uncover the greatness of Mendel's investigations.

The lack of attention to Mendel's work may also be explained by the near obsession with evolution in the mid-nineteenth century after the publication of Darwin's *The Origin of Species*. Darwin's work was published just 6 years before Mendel's and captured public attention well into the twentieth century, leaving Mendel's theory to languish quietly. (21)

The "rediscovery" of Mendel in 1900 was driven in part by what biologist Ernst Mayr calls "an accelerating interest in the problem of inheritance." (22) Incredibly, in the spring of that year three botanists—Hugo de Vries, Carl Correns, and Erich Tschermak–all claimed to have discovered laws of inheritance. They soon learned, unfortunately, that Mendel's work was nearly identical and had preceded them by 35 years. (23) In the coming decades Mendel's laws of segregation and independent assortment would be tested on a wide variety of species.

IN THE FLY ROOM

With only four pairs of chromosomes, the ability to produce offspring at a pace that would make even the most reproductively prolific blush, and the fact that it can live in the austere environment of a laboratory storage bottle, the six-legged *Drosophila melanogaster*, or fruit fly, has been the workhorse of genetics for almost 100 years. Beginning early in the twentieth century, Thomas Hunt Morgan and his students at Columbia University capitalized on *Drosophila's* valuable qualities and began breeding fruit flies by the hundreds of thousands, hoping to find variations or mutations in fruit fly traits that would help explain Mendel's laws in real-life situations. Morgan's laboratory, dominated by work with *Drosophila*, became known as the fly room, a moniker that can only partly suggest the overwhelming number of flies present in a space that measured just 16 × 23 feet. (24) Today the fly room is frequented by one of the present authors—part of it still exists at Columbia University.

During the 1910s, thanks in large part to the work conducted in the fly room, genetics shifted from simply testing Mendel's laws of inheritance to studying the physical arrangement of genes on chromosomes. Interestingly enough, the terminology of what we now call genetics was not even in place. Morgan and his genetically minded colleagues were pioneers in a field that was quickly becoming known as genetics, a word coined by botanist William Bateson in 1906. The word gene was itself first defined by the German biologist Wilhelm Johannsen in 1909. (25) The new terminology and the field of work and entity it describes are still used today.

Morgan, formerly a critic of Mendelian theory, came to embrace the new genetics because of some surprising results in his own research. In 1910 he discovered something startling among one of his breeds of *Drosophila*—a lone white-eyed male fly. When it was bred with a normal (red eyed) female, all of the offspring had red eyes. When flies from the first generation were crossed, the white-eyed character reappeared, but surprisingly only in half of the males. Finally, when white-eyed males were bred with first-generation females, 50% of both males and females had white eyes. Morgan called this change a mutation and spent much of his career studying such mutations in order to decipher the nature of genes and the structure of chromosomes. (26) Ultimately Morgan saw that Mendelian laws of segregation and independent assortment easily explained these patterns. Morgan's biographer Garland Allen suggests that these results were the main factor in Morgan's acceptance of Mendelism. (27)

The white-eyed *Drosophila* was a mutant variation of the normal red-eyed type. These types of mutations in physical characteristics became the means by which Morgan and his students at Columbia began to describe the physical entities of genes and chromosomes. People tend to think of genetic mutations as frightening, a

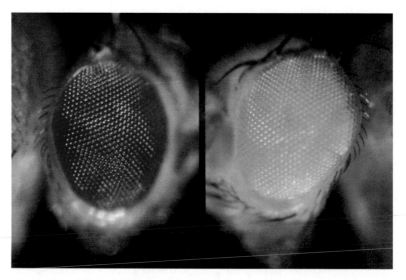

Most *Drosophila* look like the red-eyed fly on the left. Morgan's discovery
of and breeding experiments with the mutant white-eyed variety, as seen
on the right, confirmed Mendel's basic laws.

change caused by exposure to something dangerous or a freakish event or accident.
A few things drive this fear. Most obvious is a misunderstanding of what a genetic
mutation is and what it means for an organism. The other is that people have often
described mutations as the result of exposure to atomic radiation either in real life
(Hiroshima or Chernobyl) or in science fiction (Godzilla). It is true that the ill effects
on people exposed to high levels of radiation at atomic bomb sites are real and
that cancer rates among survivors of Hiroshima and Nagasaki were substantially
higher than normal because of mutations caused by the atomic bomb's radiation.
(28) But mutations are generally not of this type, nor do they create the Godzilla-like
creatures that have appeared in science fiction for the past half-century. Mutation
comes from the Latin word meaning change, so a mutation is simply a change in an
organism's DNA sequence—a change that may have no measurable effect on the
organism or may confer either a beneficial or adverse effect. Random errors that
occur during cell division are the most common cause of mutation. Most mutations
are unpredictable, as are their effects.

There are two types of mutations. One is somatic—that is, its effects die with the
organism. The other type of mutation occurs in the germline (in reproductive cells)
and can be passed between generations. But cells are resilient. During cell division
errors do occur, most of which are repaired by cellular mechanisms that are constantly at work to thwart the proliferation of cells with mutated nucleotides. During
cell division, repair mechanisms check to make sure that the correct nucleotide has

been selected at every stage of DNA synthesis. This is a tremendous task—in the human genome more than 3 billion bases are read and checked each time a cell divides. These repair systems are redundant several times over. During mammalian cell division, for example, a gene called p53 plays an important role as a cellular safety device—it can stop cells with damaged DNA from reproducing themselves. This has earned this gene the nickname "guardian angel of the genome." (29) Mutations in the p53 gene seem to play a significant role in the development of human cancers. Typically, a mutated p53 is not as effective at controlling the proliferation of cells with damaged DNA, and dangerous mutations can grow over time to become cancers.

The cause of a mutation can be the result of exposure to radiation, but as was the case in the Morgan lab, the causes of mutations for a white-eyed variation were probably far more ordinary. The white-eyed trait most likely arose from a random error in the DNA replication process. Less likely, the mutation may have been caused by a mutagen, an agent that can cause mutation. Temperature changes during gestation, environmental exposures, certain viruses, radiation, ultraviolet light, and chemicals can all act as mutagens. By using the mutations found in *Drosophila*, Morgan was able to begin to map the *Drosophila* genome. (30) This was not like the modern genome sequence maps that we hear a lot about today. Indeed, although DNA had already been isolated from cellular material, it was not yet even suspected to be the "stuff" of heredity. Thus there could be no map of the sequence of this genome, as neither science nor technology was even close to accomplishing this feat. Instead, Morgan began to map the location, or linear arrangement, of particular genes along *Drosophila* chromosomes. Working with a series of mutations, including variations in body color and wing shape, Morgan and his collaborators were able to create chromosome maps showing the location of certain genes on each of *Drosophila's* four chromosomes. (31) Morgan's group, for example, determined that the white-eyed mutation lies on the X, or *Drosophila* sex chromosome. (32)

The beauty of Morgan's work stemmed from his powers of deduction. Morgan could never actually see the positions of genes on the *Drosophila* chromosomes, but he could create virtual maps based on his experiments and deductions. Faced with unknown and unpredictable challenges neither he nor his colleagues on the genetic frontier could have anticipated, Morgan's team was able to organize information in a fashion that is as elegant and relevant today as it was when his discoveries were made. Morgan's biographer Garland Allen notes that "there have been few research groups in modern biology that have functioned as effectively together as did Morgan's group in their fly room between 1910 and 1915." (33) To develop chromosome maps, the Morgan lab used a technique that came to be known as the three-point cross. Morgan reasoned that two genes very close to each other on a chromosome would appear to stay with each other even when other parts of the chromosome recombined. By finding and generating hundreds of visible mutations, Morgan was not only able to arrange these into linkage groups on chromosomes based on whether or not they segregated together, but also to say how the traits were organized on the chromosomes. (34)

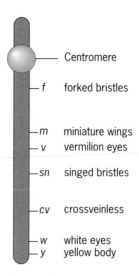

	Centromere
—f	forked bristles
—m	miniature wings
—v	vermilion eyes
—sn	singed bristles
—cv	crossveinless
—w	white eyes
—y	yellow body

Morgan's experiments with *Drosophila* led to the development of the first map of an organism's genes. This modified map shows the location of some genes on the *Drosophila* X chromosome.

But all was not well on the genetic frontier. These new and very powerful ideas concerning heredity, just beginning to make sense to some and still unknown to most, became a way to understand the world not only scientifically, but also socially.

EUGENICS—"PREVAILING SPEEDILY OVER THE LESS SUITABLE"

Morgan was not alone in his search for the mechanisms of heredity. The meanings of heredity captured the attention of natural and social scientists and, of course, the general public. While the work of Morgan and his colleagues dominated the scientific understanding of heredity during the first three decades of the twentieth century, a group of men and women known as eugenicists dominated the public understanding of heredity. These eugenicists, working under the assumption that all traits were heritable and genetic, burst onto the scene beginning in the 1890s, inspired by the work of Francis Galton in England. (35) Galton, a first cousin of Charles Darwin, defined the practice of eugenics as the science of giving "the more suitable races or strains of blood a better chance of prevailing speedily over the less suitable." (36)

The early twentieth century was a turbulent time in world history, particularly in the United States, when an influx of immigrants from Europe and the migration of African Americans out of the Deep South were challenging America's cultural and racial hierarchy. (37) Discoveries in genetics were seized on to aid in the development of social theories concerning human difference. This ultimately gave rise to eugenics, the science of improving the qualities of humanity through selective breeding. Henry Fairfield Osborn, a prominent eugenicist and president of the American Museum of Natural History from 1908 to 1933, noted that "to know the worst as well as the best in heredity; to preserve and to select the best—these are the most essential forces in the future evolution of human society." (38) "The social application of eugenic theories," one historian writes, "led to specific, detrimental effects on the lives of scores of immigrant families in the United States and to the genocide against Jews in Germany." (39)

Immigration restrictions in the United States were buoyed by eugenicist sentiment. Harry Laughlin, the superintendent of the Eugenics Record Office at the Cold Spring Harbor Laboratory, appeared before Congress several times in the early 1920s promoting his belief that immigration was foremost a "biological problem." The Cold Spring Harbor Laboratory, headed by Charles Davenport, was for all intents and purposes the headquarters of eugenics in the United States during first 40 years of the twentieth century. As Davenport's number two at the laboratory, Laughlin fervently promoted eugenics, maintaining, for example, that recent immigrants from eastern and southern Europe were afflicted "by a high degree of insanity, mental deficiency, and criminality." In his testimony before the House Committee on Immigration and Naturalization, Laughlin pleaded with Congress to restrict immigration so the United States would be allowed to "recruit and to develop our racial qualities." (40)

Sterilization laws across the United States were also inspired by eugenic sentiment. Between 1900 and 1935 approximately 30,000 so-called "feeble-minded" Americans were sterilized "in the name of eugenics." (41)

Criminals and those accused or convicted of sexual offenses were a primary concern of these eugenic laws. In 1907 the state of Indiana established the first sterilization law. By the early 1930s more than 29 other states had passed similar laws. (42) Advocates of criminal sterilization wrote that "criminals should be

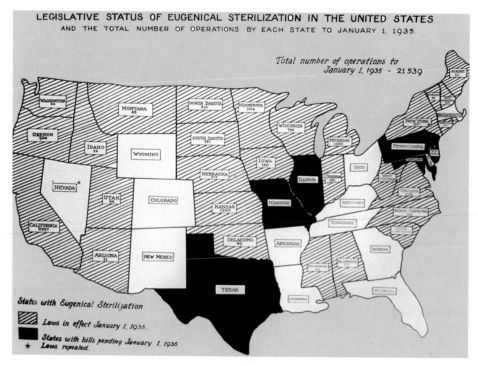

This eugenic era map shows an estimate, state by state, of the number of individuals sterilized in the United States through January 1935.

studied for evidence of dysgenic traits that are germinal in nature. Where found in serious degree parole should not be granted without sterilization." (43) "Criminality," "feeble-mindedness," and "idiocy" were all traits that could be bred out of the species—traits eugenicists believed followed Mendelian patterns of inheritance and could therefore easily be excised. (44)

On matters of race, eugenicists were also quite vocal. This period "saw the dominance of the belief that human races differed hereditarily by important mental as well as physical traits, and that crosses between widely different races were biologically harmful." (45) Well-respected geneticists wrote openly that "miscegenation can only lead to unhappiness under present social conditions and must, we believe, under any social conditions be biologically wrong." (46) In this same spirit eugenic racial science became a deviously powerful force in the Third Reich.

To a degree, Nazi eugenic zeal was inspired by American eugenics. The publication of Madison Grant's infamous eugenic tract *The Passing of the Great Race: The Racial Basis of European History* might have preceded the rise of Nazism by more than a decade, but its ideas about Nordic racial purity influenced many Germans. The book, translated into German, explicitly stated, "The laws of nature require the obliteration of the unfit." (47) In a letter to Grant, Hitler called the *The Passing* "his Bible." (48) In 1933 the *Eugenical News*, the official newsletter of several eugenic organizations including the American Eugenics Society, noted the American influence on German sterilization policy: "To one versed in the history of eugenic sterilization in America, the text of the German statute reads almost like the American model sterilization law." (49) American philanthropists, including the Rockefeller Foundation, also gave scientific grants to German eugenicist researchers both before and for several years after the rise of Hitler. (50) And even as the world recoiled in horror at the ways in which the Nazis integrated eugenics into their political philosophy—mass sterilizations, concentration camps, and gas chambers—American eugenicists continued to support their Nazi brethren. In 1935 Harry Laughlin accepted an honorary degree from the University of Heidelberg for "being one of the most important pioneers in the field of racial hygiene." The dean of the University of Heidelberg's medical school later helped organize the gassing of thousands of mentally handicapped adults. (51) Also in 1935, after a visit to Berlin, the head of the Eugenic Research Association, Clarence Campbell, proclaimed that Nazi eugenic policy "sets a pattern which other nations and other racial groups must follow if they do not wish to fall behind in their racial quality, in their racial accomplishments, and in their prospects for survival." (52) Finally, in 1937, American eugenicists distributed a Nazi eugenic propaganda film to promote the eugenic cause in the United States. (53)

After World War II outward support for eugenics became unacceptable to most biologists. The eugenic horrors of the Holocaust all but guaranteed that. And work by prominent geneticists "countered the eugenicists' simplistic assertions that complex behavioral traits are governed by simple genes." (54) But even though eugenics as an organized movement ended, eugenic ideas did not. Throughout the twentieth century ideas about heredity, social behavior, and human breeding have come in various guises, creating a fear among some that the Human Genome Project will open the door to eugenics once again.

"A VERITABLE 'AVALANCHE' OF NUCLEIC-ACID RESEARCH"

By the 1930s the ideas of Charles Darwin were once again making prominent scientific headlines. Darwin's theory of evolution lacked the mechanism to explain heredity. His theory articulated a "big picture" of evolution. He was right when he explained the ways in which evolution worked, but his theory was incomplete without genetics. Darwin's theory could not explain how evolutionary traits were passed through time. (55) Evolutionary biologists like R. A. Fisher, J. B. S. Haldane, and Sewell Wright successfully bridged the gap between evolution and genetics and spent their careers developing the mathematical framework for incorporating Mendelian genetics into evolutionary biology. This significant body of work led to what is known as the Modern Synthesis in biology. This allowed scientists like Ernst Mayr, Theodosius Dobzhansky, and George Gaylord Simpson, who were based more in data collection than in theory, to develop an empirical approach to evolutionary biology and to open up evolutionary ideas for a broader interpretation in a genetic context. (56)

While the Modern Synthesis provided a framework for understanding questions about heredity in the context of evolution, other scientists were still trying to determine the chemical components of the hereditary material. Some remained wedded to the belief that proteins transmitted traits between generations, among them Hermann Muller, who had originally worked in Thomas Hunt Morgan's laboratory, whereas others argued that nucleic acids were the fundamental elements of life. (57) No one had been able to prove this either way until a series of ingenious experiments conducted in 1944 by Oswald Avery, Maclyn McCarty and Colin MacLeod showed that nucleic acids constituted genes. (58)

Working with pneumococcal bacteria, the cause of pneumonia, Avery, McCarty, and MacLeod showed that a benign or harmless strain of pneumococci could be made virulent if mixed with dead bacteria from the same species of pneumococci that were of the virulent type. The benign strain somehow picked up the characteristics of the virulent strain and itself became a deadly form of the bacteria. Just how did this happen? How did the bacterium transform itself? Somehow, a substance in the dead virulent strain was picked up by the active strain. This "transforming principle," as it become known, altered the bacteria. To show this, the scientists isolated proteins from the virulent strain and mixed them in a laboratory culture with the benign strain. No effect was measured—the bacteria were unchanged. However, when nucleotides from the virulent strain were isolated and mixed with the benign strain, the bacterial culture turned virulent. There it was. They had purified the bacterium's proteins from its nucleic acids. DNA was the transforming material and the chemical component of genes. One biologist called the findings "electrifying" and became "convinced that it was now conclusively demonstrated that DNA was the genetic material." (59)

Every living thing on Earth—every plant and animal, every bacterium, and even viruses—shares one of the most fundamental structures of life, molecules called nucleic acids. When DNA came to be known as the stuff of heredity, focus immediately shifted from simply understanding its function to understanding its physical structure and chemical characteristics as well. Although work in this area had begun over 70 years earlier in Germany when Friedrich Miescher discovered nucleic acids in 1869, it was Avery, McCarty, and MacLeod's discovery that unleashed what one observer called a "veritable 'avalanche' of nucleic-acid research." (60) Many scientists in related fields excitedly began studying DNA,

including biochemist Erwin Chargaff, who remodeled himself as a molecular biologist and shifted his work to studying nucleic acids. This was a particularly common move among biochemists, who were well suited for DNA research because
of their training in chemistry and biology.

With DNA's structure as yet unknown, Chargaff turned his attention to the
chemical characteristics of nucleic acids. In DNA there were four known bases—
adenine, guanine, cytosine, and thymine–which are commonly referred to by their
first letters, A, G, C, and T. Each of these bases has different structures and characteristics. Analyzing the number of these bases with a chromatographic technique,
Chargaff came to a startling conclusion—in all the organisms he studied the amount
of A in any given cell was always equal to the amount of T in the same cell. The
same went for G and C. The ratio of A to T and G to C was always 1. This 1 to 1 ratio
became known as Chargaff's rule and is still one of the cornerstones of molecular
biology. (61)

Chargaff's rule
$$\%A = \%T$$
$$\%G = \%C$$
and
$$A + G = C + T$$

Many wondered how Nature could be so exact across all species on Earth. The significance of Chargaff's rule would not be entirely clear until the three-dimensional
structure of nucleic acids was determined. To do this, scientists had to take an
actual look at the physical structure of DNA, which they began to do in the 1940s.
Once they "saw DNA," the pieces of the puzzle fell into place very quickly.

A STRUCTURAL MILESTONE

Genetics in the twentieth century saw many milestones, including the work we have
already described by scientists like Morgan, Avery, and Chargaff. This work and
the work of their collaborators and colleagues propelled the revolution in genetics
forward. Their discoveries alone are striking for the ways in which they advanced
thinking in heredity. The discovery of the structure of DNA in 1953, however, has
garnered all of the headlines. On both sides of the Atlantic scientists were working
on cracking the structure of DNA. Solving this puzzle was important because it
would expose the fundamental structure of heredity and show how the molecule
at the center of life replicates itself and functions. Although chemists had already
identified the molecular components of DNA—"that nucleic acids were very large
molecules built up from smaller building blocks, the nucleotides"—James Watson
remembers that in the years preceding the discovery of DNA's structure "there was

almost nothing chemical that the geneticist could grasp at." (62) Three prominent groups worked on solving this problem: James Watson and Francis Crick at Cambridge University, Maurice Wilkins and Rosalind Franklin at King's College, London, and Linus Pauling and Robert Corey at the California Institute of Technology.

James Watson and Francis Crick are seen here at Cambridge University around the time of their discovery of the structure of DNA.

Work on unraveling the structure of DNA was most intense during 1952 and early 1953. In January 1953 Pauling's group claimed that it had solved the puzzle, proposing that DNA was a triple-stranded helix. Pauling, who had already uncovered the structure of proteins, was perhaps overzealous in his pursuit of deciphering the structure of DNA, and as a result, one of the greatest chemists in the world made an error in his calculations. (63) Scientists in England quickly picked up on Pauling's mistake. Watson and Crick recognized the error immediately as one they had almost made more than a year earlier. In the wake of this miscalculation they quickened the pace of their own research. (64)

In the early 1950s the Cavendish Laboratory at Cambridge University housed an amazing faculty of physicists, biologists, and chemists who helped create an atmosphere in which Watson and Crick could conceive of and construct models of the structure of DNA. One of the important experimental tools that Watson and Crick utilized was "pictures" of molecules. This required special physical and chemical techniques because molecules are so small. Snapshots could be taken of these extremely small molecules by first making crystals of proteins and other small molecules like nucleic acids. To take a "snapshot" of DNA, small waves of X-rays were passed through the crystals. The diffraction of these X-rays by the atoms in the DNA crystal were in essence "pictures" of these extremely small molecules. This technique, known as X-ray crystallography, allowed the scientists at Cavendish and other laboratories to interpret the three-dimensional structure for any molecule that could be crystallized.

Rosalind Franklin, a physical chemist at King's College, London, was also working on solving the structure of DNA and happened to be one of the world's leading X-ray crystallographers.

Once called the "dark lady" by her colleague Maurice Wilkins, Rosalind Franklin's valuable scientific work and her important role in the discovery of the structure of DNA have often been overlooked.

Her DNA photos were once described as "among the most beautiful X-ray photographs of any substance ever taken." (65) Just a few weeks into 1953, one of these snapshots was shown to James Watson without her knowledge or permission. Watson wrote in *The Double Helix*, his memoir of the discovery of the structure of

DNA, that "the instant I saw the picture my mouth fell open and my pulse began to race." (66) Franklin's superior X-ray crystallography enabled Watson and Crick to take the intellectual leap they had needed to complete their model of DNA.

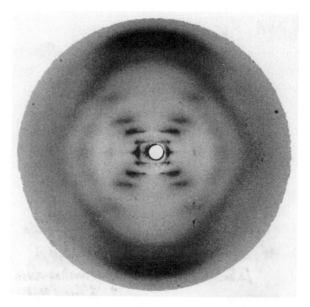

Franklin's X-ray crystallography of DNA, shown to James Watson without her knowledge, helped Watson and Crick solve the puzzle of DNA. Franklin was renowned for her X-ray crystallography talents.

Using X-ray data, including the measurements of the shape of DNA shown in Franklin's photo, Watson and Crick, piece by piece, figured out that DNA was shaped like a spiral staircase or a double helix. (67) The hereditary molecule was two chains of nucleic acids connected to one another like two snakes coiled together. The sugar backbones of the nucleotides are like supports under each step in a staircase. The nucleotide bases bond to form structures that are like steps, each one rotated slightly in relation to its neighbors in the stack. The steps that span from rail to rail of each side of the staircase are of equal length because of the specific way that two nucleotides pair. To develop their model of DNA, Watson and Crick followed Chargaff's rules closely and discovered that the double helix was complementary. That is, to form the staircase an A on one strand is always directly across from and connected to a T on the other; likewise, a G on one strand is always directly across from and connected to a C on the other. The complementary nature of the double helix revealed how DNA replicated itself and passed genetic information between generations. This process occurs during cell division when the double helix splits apart and makes identical copies of itself.

Chargaff's rules made Watson and Crick's three-dimensional model a reality. The great strength of Watson and Crick lay in their ability to reconcile their model with existing science. None of the other participants in this discovery put the pieces together quite as Watson and Crick had. And so they built their model.

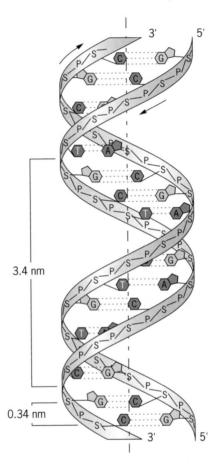

This diagram shows the double-helix structure of DNA. In the model you can see where hydrogen bonds bond the nucleic acids to one another and also the sugar-phosphate backbone that holds the helix in place.

There are three important chemical forces that hold together the DNA molecule. The first chemical bond, hydrogen bonds between a G and a C or an A and a T, connects the two strands of the helix. These bonds are relatively weak and can be broken apart by acids and/or heat. At approximately 90 °C the hydrogen bonds across a double helix can be broken, allowing the two strands of the double helix to separate.

The second kind of bond, the phosphodiester bond, keeps the Gs, As, Ts, and Cs together along a helix's strand. These bonds can be made on both ends of a base and to any other base, resulting in long strands of Gs, As, Ts, and Cs. Phosphodiester bonds are the strongmen of the helix, withstanding high temperatures and even highly acidic conditions. It is the position of these bonds on the nucleotide carbon rings that give DNA its helical twist, its third dimension. Molecular biology takes advantage of the characteristics of both hydrogen and phosphodiester bonds all the time. Because of the difference in relative strengths between these bonds, scientists would later figure out how to separate the two DNA strands. Why is this so important? Because to make copies of a double helix, you need to have both strands—let's call them the Watson and Crick strands—as a template. If they are bonded in a double helix, they cannot be used to replicate themselves. By melting the weak hydrogen bonds between the two strands, the freed strands can now be copied.

The race to uncover the structure of DNA became the stuff of scientific legend after the publication in 1968 of James Watson's *The Double Helix*. Watson's telling of the DNA story drew ire from within the small community of scientists in which he himself worked. Facing strong objections from Francis Crick, Linus Pauling, Maurice Wilkins, and the family of Rosalind Franklin over the way in which Watson characterized all of the major players in the discovery, Harvard University Press dropped the book. (68) Of particular concern was Watson's portrayal of Franklin, who was just thirty-seven when she died in 1958 of ovarian cancer and whose role in the discovery was reduced in *The Double Helix* to that of an incompetent scientist and hot-tempered woman. (69) Watson's book, picked up by another publisher, went on to become a best-seller, and for almost 30 years the story of the discovery of DNA was told by *The Double Helix*. It is only recently, with the publication of a new biography and with acknowledgments by Watson that Franklin's work was "key" to their success, that Franklin's image as a brilliant scientist was rehabilitated. Years after his own codiscovery of the structure of the double helix, Francis Crick suggested that Franklin was just months away from solving the puzzle herself. (70)

As late as 1933, Thomas Hunt Morgan suggested that there is "no consensus opinion amongst geneticists as to what the genes are—whether they are real or purely fictitious." (71) Working deductively, working on instinct, Morgan could never be sure that his gene maps or the work on genes conducted by his many colleagues amounted to anything. But beginning with Avery, McCarty, and MacLeod's discovery in 1943 that DNA was the "stuff" of heredity, the gene became less an intellectual or theoretical entity and more a material reality. Watson and Crick's discovery of the actual physical structure of DNA finally created a consensus among geneticists that genes were real and led genetics and molecular biology into a new and exciting realm. With the basics of heredity worked out, molecular biology became a driving force in science as the working characteristics of the gene came under scrutiny and study.

The Building Blocks of Gene Sequencing

In the 1940s the Nobel Prize-winning physicist Erwin Schrödinger inspired a generation of scientists to study genes. Known primarily for his work in quantum mechanics, Schrödinger spent World War II in exile in Dublin, where in February 1943 he gave a pioneering series of lectures at Trinity College on the importance of understanding the physical laws that govern heredity. These groundbreaking talks, published a year later as a slim volume entitled *What Is Life?*, anticipated the importance of DNA just as scientists began to establish the nature of what Schrödinger called "the most essential part of a living cell—the chromosome fibre." (1) With no formal training in the natural sciences, Schrödinger came to genetics with what he called "a naïve physicist's ideas about organisms." (2) Despite this limitation, by combining a physicist's sense of the need for order in the natural world with a sophisticated understanding of contemporary currents in the biological sciences, Schrödinger was able to articulate a prescient and stirring vision of what was to come in molecular biology.

Schrödinger's call to biologists posed the central question of *What Is Life?*:

> How can the events in space and time which take place within the spatial boundary of a living organism be accounted for by physics and chemistry? (3)

The almost 90-page answer to this question, considered by many to be one of the masterpieces of scientific literature, speculated on how hereditary material survives in conditions outside of the known boundaries of physics to pass on genetic information generation after generation. He suggested that the answer lay in the chromosome fiber, which, compared with the "rather plain and dull" material of inanimate nature, was more like a "masterpiece of embroidery," "an elaborate, coherent, meaningful design traced by the great master." (4) This, the most basic idea in *What Is Life?*, influenced the young generation of biologists in the 1940s

27

Welcome to the Genome, by Rob DeSalle and Michael Yudell.
ISBN: 0-471-45331-5 Copyright © 2005 Rob DeSalle and Michael Yudell.

and early 1950s to follow Schrödinger's clarion call and search for the rules that underlie genetic matter. Schrödinger's influence is unmistakable. After reading *What Is Life?* James Watson remembered that he "became polarized towards finding out the secret of the gene." Francis Crick, Watson's codiscoverer of the structure of DNA, recalled the impact of the book, remembering that it "suggested that biological problems could be thought about in physical terms—and thus it gave the impression that exciting things in this field were not far off." (5)

Schrödinger's reduction of life to the laws of physics and chemistry need not be read as a deterministic view of the primacy of genetic heredity over the other factors that determine an individual (i.e., the components of a person's environment). After all, the question of his book is "What is life?" and not "What makes us human?" or "What is the meaning of life?" Instead, Schrödinger was after something much more basic—the substances and rules that determine genetic heredity—that from a physicist's viewpoint was essential to understanding life. The discoveries discussed in this chapter reflect Schrödinger's conviction that the substance of life can be reduced to interplay between physics and chemistry. Yet, although these discoveries illustrate the mechanistic nature of genetic heredity, they cannot paint a complete picture of why we are the way we are. "The answer to *What is Life?*" the evolutionary biologist Stephen Jay Gould reminds us, "requires attention to more things on earth than are dreamed of in Schrödinger's philosophy." (6)

This chapter examines some of the essential components of the gene sequencing puzzle and of the growing general understanding of the mechanisms of heredity. Today we can look back on these discoveries and see how they are like stations along an assembly line, making up separate pieces that are all essential to the overall product of gene sequencing. In Chapter 3 we will see how all of these technologies came together to give us the technology that sequenced the human genome.

COMPONENT 1: BASIC SEQUENCING TOOLS

Nearly 50 years passed between the discovery of the double helix and the sequencing of the human genome. Some of the earliest techniques developed by scientists working on the problems of genetic heredity so closely resemble methods used by contemporary genome scientists that it may seem surprising that it took so long to complete the human gene sequence. But molecular biology was still in its infancy in the 1950s, and the technological advances necessary to sequence a whole genome would still take decades to come to fruition.

The first big step forward for sequencing technology took place at Cambridge University, England, in the mid-1950s in the laboratory of biologist Frederick Sanger. Well before gene sequences, in the earliest stages of our understanding of how genes function, Sanger discovered how to take a protein, break it down into its component parts, and, piecing the puzzle back together, determine the order of amino acids along a protein. His ingenious approach to understanding the sequencing of proteins eventually won him his first of two Nobel Prizes and was the conceptual precursor to contemporary DNA sequencing. (7) An understanding of proteins was also important because of the role these complex molecules play in an organism. Proteins receive their instructions from genes to carry out such diverse tasks as food digestion, production of energy in a cell, transmission of impulses in the nervous system, and the ability to smell, see, and hear. If genes and DNA are the material that perpetuate heredity and help determine an organism's form and function, then proteins are the cell's workhorses, carrying out the varied instructions inscribed in an individual's DNA. Proteins can also play a harmful role in an organism. Genetic defects can cause the absence or overabundance of a particular protein, which in both cases can cause devastating illnesses. For example, phenylketonuria, or PKU, is a metabolic disease caused by a genetic defect that leaves individuals without a protein that breaks down the amino acid phenylalanine. A buildup of phenylalanine causes severe mental retardation. Individuals diagnosed with the disease as infants can alter their diets to keep levels of phenylalanine low and avoid PKU's dreadful effects. (8)

The method developed by Sanger exploited the chemistry of amino acids and proteins that had been well known for over 10 years. Just as nucleotides are the building blocks of DNA, amino acids are the building blocks of proteins. Sanger himself wrote in the journal *Science:* "In 1943 the basic principles of protein chemistry were firmly established. It was known that all proteins were built up from amino acid residues bound together by peptide bonds to form long polypeptide chains. Twenty different amino acids are found in most mammalian proteins, and by analytical procedures it was possible to say with reasonable accuracy how many residues of each one was present in a given protein." (9)

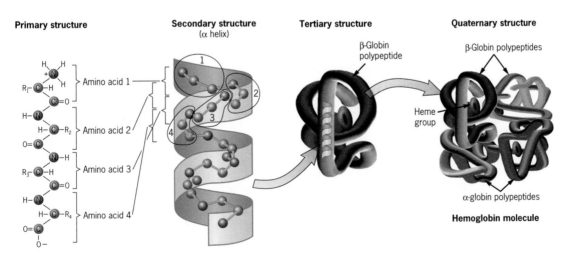

This figure shows the way in which amino acids are the building blocks of proteins. In this case, we can see how a hemoglobin molecule is made up of a string of amino acids.

Sanger's challenge was to figure out a way to read the order of the amino acids that determine a protein. For his experiments Sanger chose to use bovine, or cow, insulin because of its important medical significance and its relatively short length—only 105 amino acids. Sanger set out to find ways to read the unwieldy molecule, which by his method could be deciphered only by breaking the protein apart, looking at small stretches of four or five amino acids, and then conceptually putting the molecule back together like a puzzle to determine the full sequence.

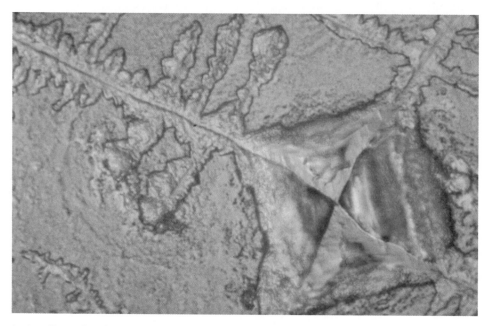

An insulin molecule is shown here in crystal form.

Sanger determined that the exposure of insulin to certain chemicals could break the peptide bonds in a protein chain. Sanger was able to identify the kinds of amino acids these broken-down parts contained. He then created groups of small chains of amino acids that could be "tiled," or pieced together, to give a full-length sequence of a protein. (10)

Sanger was considered to be "reticent, even shy, a man who worked with his hands, at the laboratory bench." (11) Yet he also recognized the impact that his work would have on science and medicine.

In his address to the Nobel committee in 1958 Sanger underlined the importance of understanding the chemical nature of proteins. "These studies are aimed," he said, "at determining the exact chemical structure of the many proteins that go to make up living matter and hence understand how these proteins perform their specific functions on which the processes of life depend." He also hoped that his

Frederick Sanger played a critical role in the development of molecular biology and in the technologies that enabled the sequencing of the human genome.

work "may reveal changes that take place in disease, and that our efforts may be of more practical use to humanity." (12) This connection between proteins, genes, and medicine, uncovered in part by Sanger and his techniques, is at the heart of what lies ahead in genomics.

COMPONENT 2: THE ABILITY TO READ THE GENETIC CODE

The most basic mechanisms and building blocks of heredity were, by the late 1950s, either solved or theoretically understood. But the link between genes and proteins was still not fully established. After all, nobody had yet explained exactly how DNA could produce a protein. The growing awareness that proteins were linear arrangements of amino acids and that genes were linear arrangements of nucleotides suggested to many scientists that this could mean only one thing—there was some code that connected the information in DNA to the production of proteins. But this was no simple code to crack, and scientists had been working on variations of this problem for at least a decade before the discovery of the structure of the double helix.

The intellectual spark that was a foundation for the solution of the DNA/protein code came from an unlikely source. Soon after the 1953 publication in *Nature* of their famous paper on the structure of DNA, Watson and Crick received a letter from George Gamow, a theoretical physicist and one of the architects of the big bang theory of the universe. Gamow's letter sketched out an explanation for how an array of nucleic acids determined an array of amino acids. Gamow's model, which detailed a list of 25 amino acids, turned out to be wrong. Paring down Gamow's list to 20, Watson and Crick came up with the correct number of amino acids that make up proteins. (13) Over the next decade scientists conducted experiments that confirmed Watson and Crick's list of amino acids and uncovered the DNA/protein coding scheme.

The 20 Amino Acids

Amino acid	Three-letter symbol	One-letter symbol
Alanine	Ala	A
Arginine	Arg	R
Asparagine	Asn	N
Aspartic acid	Asp	D
Cysteine	Cys	C
Glutamic acid	Glu	E
Glutamine	Gln	Q
Glycine	Gly	G
Histidine	His	H
Isoleucine	Lle	I
Leucine	Lev	L
Lysine	Lys	K
Methionine	Met	M
Phenylalanine	Phe	F
Proline	Pro	P
Serine	Ser	S
Threonine	Thr	T
Tryptophan	Trp	W
Tryosine	Tyr	Y
Valine	Val	V

These are the 20 amino acids found in proteins. They are usually described by their three-letter abbreviations or by a one-letter symbol.

In DNA there are four linearly arranged nucleic acids (G, A, T, and C), whereas proteins are constructed from 20 linearly arranged amino acids. It was apparent from basic mathematics that the code was not based on a 1:1 relationship—the connection between DNA and proteins was not one nucleic acid to one amino acid (it would require at least 20 different nucleic acids to make a 1:1 ratio work). The code could also not be solved based on a 2 to 1 ratio. That is because there are only 16 ways G, A, T, and C can be arranged.

It turned out that the code is based on a 3:1 relationship and is therefore a series of nonoverlapping triplets of nucleic acids that code for single amino acids. Basic mathematics shows that there are 64 different ways to arrange four different bases in triplets. But there are only 20 types of amino acids. This is because some of the triplets, which are called codons, are redundant; they are just different ways to code for the same amino acid. Most amino acids have either two or four synonymous codons, although there are several exceptions. The amino acids methionine and tryptophan have no synonymous codons. Isoleucine has three, and serine, arginine, and leucine all have six.

Deciphering the genetic code allowed scientists to scan stretches of DNA sequences and look for genes. The language spelled out by nucleic and amino acids has rules similar to the rules of punctuation. Just as you can scan this paragraph for capital letters and periods, you can look for the first word in a DNA sentence to find what is called an initiator codon and read on until you find the end of the sentence or period, which in genetic terminology is called the terminator codon. Everything between these points is part of the same gene.

In a genetic sentence the initiator codon is almost always a triplet of the nucleic acids A, T, and G, which codes for the amino acid methionine (also known as Met or M). Thus, when you look at the amino acids that make up proteins, you will, with a few exceptions, always see an M as the first letter in the protein. Experiments by Cambridge University biologists Sydney Brenner and Francis Crick and by Alan Garen at Yale University showed that there were three terminator codons or three ways to put a period at the end of a protein sentence—TAG, TAA, TGA. (14)

A Sample Genetic Sentence

ATG (initiator codon) GCA AGT TCT T...GC ATA AGT TAG (terminator codon)

This sounds easier than it actually is, however. As with the English language, a capital letter does not always indicate the beginning of a sentence. Once an ATG is located, scientists must determine whether the suspected gene is actually a gene at all. The suspected gene is called an open reading frame (ORF) and this process is called annotation.

It took nearly a decade of work for experiments to confirm the triplet model of protein synthesis. In 1961 at the U.S. National Institutes of Health biochemists Johann Heinrich Matthaei and Marshall Nirenberg verified the first word of the genetic code. Matthaei and Nirenberg's experiment was relatively simple. In a test tube, they provoked nucleic acids they had synthesized in test tubes to produce a protein. Placing only one type of nucleic acid, all T's, into a test tube, they were able to produce the protein made up of only the amino acid phenylalanine, or P, meaning that the triplet TTT coded for phenylalanine. (15) Later that year at New York University School of Medicine biochemist Severo Ochoa began similar experiments constructing random strings of nucleotides, placing them in cell extracts, and determining the kind of amino acids that were incorporated into the subsequent protein. (16) By comparing the results of these and other experiments, scientists cracked the entire code of triplets by 1965.

Breaking the genetic code alone couldn't explain the relationship between genes and proteins. By the late 1950s scientists recognized that some type of intracellular intermediary was bringing genetic information from DNA to ribosomes, which are the cellular mechanisms that assemble proteins. The link between DNA and proteins turned out to be a cellular material known as ribonucleic acid or RNA. (17)

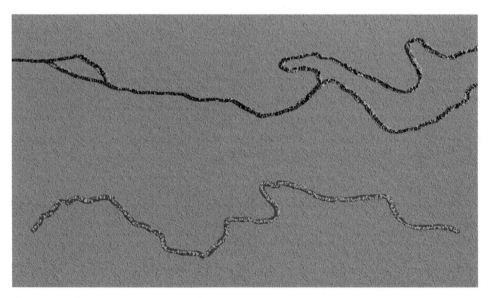

This image of DNA (top) and RNA (bottom) was taken by an electron microscope.

RNA is a versatile molecule; it acts as structural scaffolding, as an enzyme, and as a messenger. Its general structure is the same as that of DNA, but its sugar ring is slightly different, hence the deoxyribo- in DNA and just plain ribo- in RNA. Also, like DNA, RNA has four kinds of bases. However, instead of T, or thymine, RNA has U, or uracil, which complements A when RNA binds to DNA.

There are two steps in translating genetic instructions into a protein. The first is called transcription. RNA molecules assemble along a stretch of DNA that constitutes a gene. The strand of RNA is complementary to the strand of DNA by the same rules that dictate the formation of a double helix.

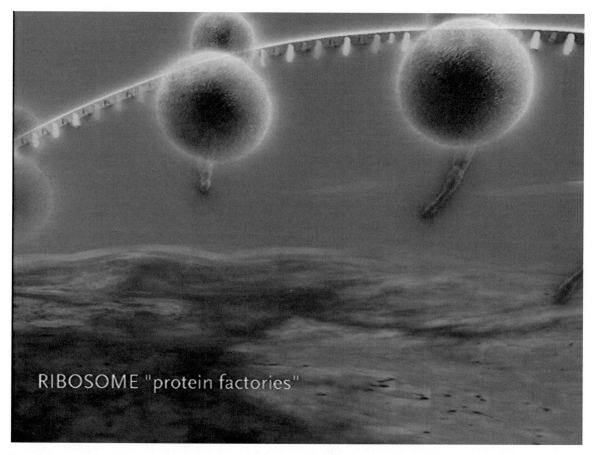

RIBOSOME "protein factories"

Proteins are made in two steps. Messenger RNA first assembles along a gene (transcription). The mRNA molecule then moves out of the nucleus to a ribosome (pictured here), where it is translated into a protein (translation).

Once formed, this strand of RNA, known as messenger RNA, or mRNA, moves out of the nucleus of a cell to a ribosome, where the genetic sentence is read and translated into a protein. This stage in protein formation is known as translation. This molecular mystery was solved by some of the same scientists working on decoding the genetic code—Sydney Brenner at Cambridge, François Jacob and Jacques Monod at the Institute Pasteur in Paris, and Matthew Messelson at Cal Tech. (18) The breaking of the genetic code allowed scientists to interpret DNA information by providing them with an accurate DNA to protein dictionary. This innovation was an important component of the assembly line of technologies that eventually shaped gene sequencing.

COMPONENT 3: MAKING DNA

Since the early part of the twentieth century, scientists had been aware of the vital connection between genes and enzymes, a type of protein that usually accelerates chemical reactions in an organism. As early as 1901, Archibald Garrod, a London physician studying metabolic disorders, recognized that patients with the disease alkaptonuria were lacking what he called a "special enzyme" that results in the body's inability to break down a substance called alkapton. By studying familial patterns of this disease, Garrod came to infer that the missing enzyme was a problem of inheritance; most of the children with the defect were born to parents who were first cousins. (19) This "shallow" gene pool made the emergence of this recessive trait more likely.

Four decades later at Stanford University, biochemist Edward Tatum and geneticist George Beadle refined Garrod's observations, suggesting in 1941 that one gene codes for one enzyme, a theory that was a cornerstone of molecular biology for more than five decades. They were awarded a Nobel Prize for their discovery in 1958. (20) Although DNA itself was coming to be known to be the stuff of heredity, enzymes and other proteins, it was turning out, were essential to the successful operation of the cell and therefore of the organism. If hereditary information was carried on DNA, then the different classes of proteins are, in large part, heredity's workhorses, delivering instructions for many of life's intricacies at the beck and call of the DNA molecule itself.

Work at the cellular level, with its varied goals, was less directed, for example, than the search for the structure of DNA. Some scientists were busy taking the cell apart to determine how DNA replicated, others learning how proteins were synthesized, and still others inquiring about the nature and function of proteins. In fact, Arthur Kornberg carried out his Nobel Prize-winning discovery of the protein in bacteria that controls DNA replication without Watson and Crick's work in mind. Perhaps what Kornberg himself called his "many love affairs with enzymes" distracted him from the broader goings-on in molecular biology. "The significance of the double helix did not intrude into my work until 1956," Kornberg wrote, "after the enzyme that assembles the nucleotide building blocks into a DNA chain was already in hand." (21)

Kornberg's discovery, once known as DNA polymerase or Kornberg's enzyme and now known as DNA polymerase I, catalyzes the addition of nucleotides to a chain of DNA (other DNA polymerases were discovered later, and were in turn known as polymerases II, III, etc.). In other words, DNA polymerase is the mechanism by which DNA clones or copies itself. Working with the bacteria *E. coli*, a bacteria that is usually beneficial to the function of the human digestive tract, Kornberg showed that the enzyme DNA polymerase was able to synthesize a copy of one strand of DNA. With a single strand of DNA in a test tube, the presence of DNA polymerase served as the catalyst (or initiator) for DNA replication. These experiments revealed only that the synthesized DNA was true to Chargaff's rules, having the correct ratio of As to Ts and Cs to Gs. (22) Kornberg's results did not, however, reveal the sequential arrangement of nucleotides, nor was it known at this time whether this laboratory model was what actually happened in living organisms. (23)

It later turned out that Kornberg's polymerase was not the key polymerase in DNA replication; DNA polymerase III was. Scientists who questioned the function of Kornberg's polymerase in live organisms were only partially correct; polymerase I's role was still found to be vital, playing a key role in chromosome replication and DNA repair. (24) Over the next two decades the approaches pioneered by Kornberg and his associates resulted in the discovery of a broad array of enzymes and other proteins important in the replication of DNA and the translation of proteins. An intriguing aspect of these discoveries is that polymerase enzymes do not need to be in cells to work. Biochemists used this feature of polymerase to develop methods to take proteins out of cells and coax them to activate in test tubes. The other enormously important result of Kornberg's work was that scientists now had a laboratory reagent–the DNA polymerase itself–that could be used in a test tube to replicate DNA.

COMPONENT 4: SEEING GENES

Sanger's sequencing of insulin's amino acids, the cracking of the genetic code, and Kornberg's work on DNA polymerase were all technologies that would someday lead to the sequencing of a whole genome. But the ever-increasing knowledge of the molecular basis of inheritance could not reach its full potential for both scientific and biomedical research without techniques to sequence genes quickly and accurately. So we now turn from deciphering the interiors of the cell to technologies that capitalized on these discoveries and enhanced our ability to see the most fundamental mechanisms of heredity. By the 1970s laboratories around the globe were focused on finding ways to better see genes and their component parts.

Oxford University biologist Edward Southern revolutionized molecular biology in 1975 with a method that came to be known as the Southern blot. (25) Southern blots allowed geneticists to locate and look at DNA and genes within a genome by capitalizing on the following characteristics of DNA. First, DNA is a negatively charged molecule; thus when electricity is present, it can hitch a ride on a current— it migrates to the positive terminal in an electric field. Second, DNA molecules are small and can be separated by passing them through a porous gel made from either agarose (extracted from seaweed) or acrylamide (a synthetic polymer). The size of the DNA fragment, the strength of the current, and the concentration of acrylamide or agarose in the gel mixture dictate how fast molecules will pass through it. In fact, the concentration of an acrylamide gel can be adjusted to such a fine degree that DNA molecules of one base pair difference in length can be distinguished. Third, one fragment of DNA can be used to find another. This process, known as DNA hybridization, activates one strand of a double helix to search for the other strand, to reform hydrogen bonds and make a new double helix.

DNA hybridization occurs when a single strand of a double helix finds a complementary strand to form a new double helix.

Hybridization doesn't have to be perfect; only 60%–70% of the two strands of a helix must match for the two strands to stick together.

Southern created a technique whereby a small piece of an organism's genome can be arranged by size along an agarose gel. The technique then involves the transfer of DNA from gels onto nitrocellulose membranes. To detect where a specific gene is, a fragment of the gene of interest is labeled with radioactivity and then hybridized (attached) to the DNA on the nitrocellulose. Radioactive molecules hybridizing to the DNA on the filter emit particles that react with photographic film and can therefore be seen as a dark spot on the film. With his blot, Southern solved the problem of finding a genetic needle in a genomic haystack. Techniques were later developed to similarly isolate RNA molecules (jokingly named Northern blotting) and protein molecules (named Western blotting). (26)

COMPONENT 5: COPYING DNA

During the 1970s scientists improved upon the Southern blot and other gel electrophoresis methods. Southern's method required a tremendous amount of DNA and thus a tremendous amount of laboratory labor. It also lacked the precision to see the location of individual bases. To get around this shortcoming, scientists developed methods to amplify or clone (meaning simply to copy) DNA.

Found in bacteria, the small, circular extrachromosomal DNA molecule known as a plasmid often times carries genes that confer antibiotic resistance to bacteria. Under normal conditions a plasmid can facilitate genetic exchange between different bacterial strains: A gene fragment from one bacterium is carried to and inserted into a chromosome by a plasmid. In 1973 scientists discovered that if you biochemically insert a target stretch of DNA into a plasmid and put the plasmid into a bacterial cell, such a cell makes thousands and perhaps millions of copies of the plasmid and hence the attached DNA. (27) This procedure, incorporated into sequencing technology, made it easier to make large amounts of a desired stretch of DNA. There are, however, two serious shortcomings of the use of plasmids. First, bacterial plasmids must be cultured. This is time consuming. Second, plasmids can take up only a small piece of DNA efficiently. If the DNA stretch picked up by a plasmid is too large, the plasmid is unable to make accurate copies.

Since the discovery of plasmids or what might be termed bacterial copying machines, other vehicles have been created that can copy larger pieces of DNA. The average limiting size of a plasmid is about 5000 bases. Phages, a specific class of viruses that infect bacteria and can be stably replicated by them, can carry about 15,000 bases; cosmids, an artificial cloning vector with a phage gene, can carry about 35,000 bases; bacterial artificial chromosomes (also known as BACs) can take over 100,000 bases of sequence; and yeast artificial chromosomes (also known as YACs) can take approximately 1,000,000. Although these microbial methods remain an important component of DNA sequencing and were central to the effort to sequence the human genome, they are all arduous ways to copy DNA. BAC-copied DNA was used in the sequencing of the human genome. (28)

COMPONENT 6: MORE SEQUENCING SUCCESSES

By the 1970s advances in sequencing technology brought biology and genetics to the brink of the genomic revolution. The most important developments in sequencing technology occurred simultaneously in laboratories on opposite sides of the Atlantic. Two groups—biologist Walter Gilbert's group at Harvard and Frederick Sanger's group at Cambridge—exploited the chemistry of nucleic acids to come to the same brilliant idea. Unlike Edward Southern's method, which revealed only the presence of DNA and genes, Gilbert's and Sanger's methods revealed the actual sequences of nucleotides along strands of DNA. The two methods had "complementary strengths," and were used depending on what was to be sequenced. (29) The men shared the Nobel Prize in 1980 for this work. It was Sanger's second Nobel. (30)

Sanger's sequencing success rested on several premises. First, he knew that he could take a piece of DNA and synthesize its entire length with DNA polymerase. He was also aware of discoveries that showed that by using a class of nucleotides called chain terminators he could interrupt the synthesis of a DNA chain. These chain terminators come in four forms—terminator G, terminator C, terminator T, and terminator A—and when they were placed in a test tube with a DNA fragment and DNA polymerase and then placed on a gel, Sanger could determine the order of nucleotides in a given DNA fragment. He accomplished this by radioactively labeling the locations where the chain terminators stopped DNA synthesis at one of the four particular nucleotides. (31)

Sanger's method of labeling fragments of DNA with radioactivity, using gel electrophoresis to separate the fragments, and using X-ray film to visualize them, quickly became commonplace in molecular biology laboratories and is still today the basis for gene sequencing. (32) In 1977, using his own method, Sanger himself accomplished the once unthinkable by completing the sequence of the entire genome of Phi-X174, a virus that infects *E. coli* in the human digestive tract. Despite the fact that this virus was just over 5000 base pairs long, it took Sanger's group years to sequence it. (33) By 2000 the Phi-X174 genome could be sequenced in just a few hours.

The sequence itself revealed remarkable information about genes and gene structure. Among the most intriguing was the finding that even though there are 5386 nucleotides and nine proteins made from genes in the genome of Phi-X174, calculations showed that there was not enough DNA to code for the proteins that the Phi-X174 genome produced. This was confusing to scientists. The larger number of proteins than available DNA in PhiX was accounted for by some stretches of the genes in the PhiX genome coding for two or more different proteins by having one gene embedded in another. (34) This important finding is characteristic of many genomes, including the human genome. (35)

COMPONENT 7: THE ULTIMATE DNA COPYING TOOL

Few scientists have a moment of inspiration like the one that came to Kerry Mullis in 1983. According to Mullis, he was driving along a winding moonlit California mountain road when he thought up "a process that could make unlimited numbers of copies of genes." As he drove, he designed the polymerase chain reaction (PCR) in his head. (36) PCR would soon become the newest and most advanced gene amplification technique, allowing for millions of copies of selected fragments of DNA to be made without plasmid cloning in as little as an hour, as opposed to the tedious vector-based cloning that could take weeks or even months.

Mullis and his colleagues reasoned that four things were needed to make DNA: (1) a template (one of the strands of the target sequence from a double helix), (2) the nucleotides (the basic building blocks of DNA—G, A, T, and C), (3) primers (short single strands of DNA designed to find their base pair complements), and (4) an enzyme, a DNA polymerase. They also recognized the key from previous work on DNA replication—that in order to replicate a specific region of DNA in a genome you would need to have two primers, one for each strand to be read in opposite directions. The distance between these primers would define the length of the sequence that this new method would amplify. (37)

With these basic tools and a simple but ingenious algorithm, Mullis created the three-stage polymerase chain reaction. In the first step the temperature of DNA is raised to above 95–97 °C, a temperature at which the strands of a double helix come apart. Second, the temperature is lowered to 45–65 °C, which forces the primers to anneal or stick to the target region of DNA. Finally, at a temperature conducive to the DNA polymerase, the reaction is activated and replication begins.

These PCR cycles produce an astounding number of fragments between the two primers. It starts with only a single copy of a fragment. After one cycle of PCR two copies of the desired fragment exist. After two cycles four copies exist, and after three cycles eight copies exist. Four cycles make 16 copies. The increase in copy number is not linear, but rather geometric. Finally, after approximately 30 cycles, over a billion copies of a particular DNA segment will exist in the reaction.

Mullis had one serious problem to overcome. At 95 °C almost all cellular material denatures, destroying the needed polymerase in the PCR reaction. In the original PCR design, fresh polymerase had to be added after each cycle. By 1988, however, the cycle was modified by the addition of a DNA polymerase from the bacterium*Thermus aquaticus,* which normally thrives in and around deep-water thermal vents and easily resists the 95 °C melting temperature in the PCR cycles. The cycle could thus run continuously without adding fresh polymerase by starting it at 94 °C (denaturing the DNA strands), lowering it to 45–65 °C (to anneal the primers), and then raising it to 72 °C (to activate the *Thermus aquaticus,* or *Taq,* polymerase). (38)

PCR Can Copy DNA at an Amazing Rate

Cycle	Copies of DNA
1	2
2	4
3	8
4	16
5	32
6	64
7	128
8	254
9	508
10	1016
15	32,768
20	1,048,576
30	1,073,732,608

The PCR process rapidly copies DNA fragments. The average PCR machine can produce billions of copies of DNA in just a few hours. Newer PCR machines can produce billions of copies in as little as 30 minutes.

The molecular revolution was just over 30 years old by the mid-1980s. Although so much had been accomplished since Watson and Crick's groundbreaking discovery in 1953, the broader application of genetics was limited by the then-current state of technology. Molecular biologists had established the basic physical and chemical rules of heredity, providing the biochemical tools to answer Schrödinger's question *What is Life?* From Sanger's basic sequencing tools to the cracking of the genetic code to the development of PCR, technologies were developed that brought science closer to answering Schrödinger's question. But even with these tools scientists were only barely able to apply knowledge of cellular "life" to basic medical challenges. The genetics of sickle-cell anemia, for example, have been understood for 40 years yet there is still no cure for this disease. The proposal to sequence the human genome in 1985 was an attempt to provide biology with something akin to chemistry's periodic table. Such a catalog of the human genome, scientists hoped, would provide a foundation for improving our understanding of the relationship between genetics and human disease, and be a way to begin to apply nearly a century of work in genetics to health care. Much as Schrödinger's question prompted a generation of scientists to investigate and uncover the molecular mechanisms of heredity, the sequencing of the human genome inspired scientists at the dawn of the twenty-first century to develop a more precise and richer understanding of how our genomes work.

Sequencing the Genome

Twentieth-century biology ended a lot like it had begun—with a major milestone. At the dawn of the century three botanists independently rediscovered Mendel's laws of heredity, setting the stage for a century of genetic discoveries. At the century's end scientists from around the globe completed a draft sequence of the human genome, providing the foundation for a new era of work in genetics. So much changed in just one hundred years. A science that started the century relying on keen observational skills and what now seem like crude technologies ended it by celebrating incredible technological achievement and an ability to do what was once only imagined. The completion of the human genome sequence is a remarkable story of human ingenuity and zeal, great technological leaps, and intense competition. The race to sequence the human genome is a story in itself; books have already been written about it. (1) This chapter looks at the technologies and personalities involved in making the sequencing of the genome possible.

FROM GENE TO GENOME

Before we conclude our exploration of the age of genomic information with a look at the race to sequence the human genome, it is important to be more familiar with the relationship between genes and genomes. It is a relationship that we have been building toward by highlighting the discoveries in genetics that made sequencing the genome possible. In Chapter 1 we provided the fundamentals of the genetic lexicon and explained that, whereas genes are the basic unit of heredity in all living beings on Earth, genomes are the entire set of an organism's genetic material. This relationship is not simply one of degree. Genes and genomes don't exist independently of one another. They are part of the same hereditary structure that makes life possible.

Our bodies are made up of trillions of cells. Every tissue from muscle to nerve has specialized cells that enable us to move, think, reproduce, see, and grow. Cells vary in shape and size but each has a nucleus that contains an individual's genome

43

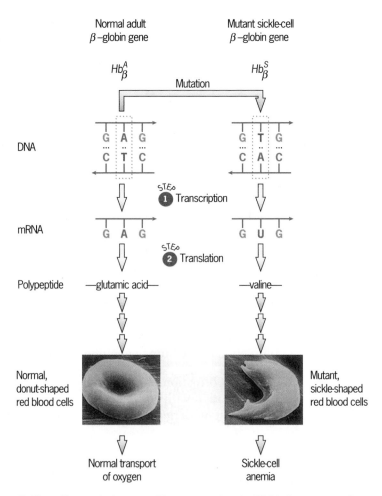

Sickle-cell anemia is caused by a mutation in DNA that causes the production of the wrong protein, resulting in sickle-shaped red blood cells.

and therein a complete set of instructions directing cell development and function. The human body produces billions of new cells each day. The approximately 30,000 genes in your genome direct this cell production. Only a percentage of these genes are expressed in any given cell. Some traits in organisms are relatively simple with respect to genes, controlled or coded for by a single gene or just a few. Among these traits are sickle-cell anemia and some forms of color blindness. Most traits, however, are very complex genetically, sometimes the result of perhaps hundreds of genes interacting among themselves and with the environment to produce a particular outcome.

The disease sickle-cell anemia is triggered by a single mutation. Along one of the genes responsible for developing normal hemoglobin, a single DNA base pair substitution causes the base adenine to replace thymine. This error leads to the production of the wrong amino acid, which in turn causes the hemoglobin molecule to be deformed and shaped like a sickle. (2)

Red-green color blindness is, like sickle-cell anemia, a relatively simple trait that can be linked to simple genetic flaws. Human vision is trichromatic, which means that our eyes can see almost any shade of color when red, green, and blue are mixed. In a normal eye, color vision occurs when light hits the retina, the light-sensing part of an eye. The retina contains two types of specialized cells—rod cells, which enable our eyes to see bright and dim light, and cone cells, which enable our eyes to interpret different wavelengths of light. In the 7 million cone cells in a human eye, red-, green-, and blue-sensitive pigments are responsible for the colors we see. (3) An error in the genes that produce any one of these pigments can cause color blindness.

A form of red-green color blindness known as deuteranopia is caused by a missing gene on the X chromosome and results in vision that sees shades of yellowish

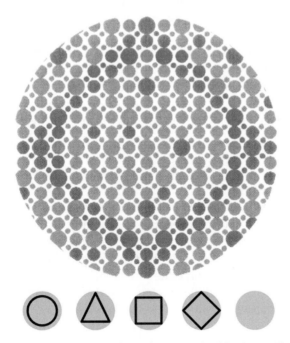

This test helps screen for red-green color blindness. If you don't see a pattern within the image on this page, chances are that your eyes cannot accurately process the colors red and green.

brown instead of distinct red and green colors. Because of its origins on the X chromosome and because males have only one X chromosome, men are far more likely to inherit this disorder. Almost 1 in 10 men of northern European extraction are red-green color-blind, whereas only 1 in 250 women have the disorder. The reason is that females have two X chromosomes and thus two chances of inheriting a chromosome without the red-green defect. Women who have a working combination of the red-green genes and who also carry the defect are called carriers. Their vision is not affected by the disorder. (4)

It is in the nuclei of cone cells that we can locate the genetic fault that leads to red-green color blindness. There is very little difference between the DNA in the nucleus of cone cells and the DNA in all the other cells in the human body. The same genetic error that causes deuteranopia is, after all, found in all cells. However, different genes are active in different types of cells according to function, and therefore this error becomes a problem only in cone cells. There are two genes involved in this disorder—the green opsin gene and the red opsin gene. Both of these genes direct cone cells to produce proteins called red and green opsins that respond to light stimuli. The green opsin protein absorbs the green spectrum of light and transmits an impulse to the brain telling the brain that the color observed is green or some shade of green. The same is true for the red opsin gene. A cone cell needs both red and green opsin in order to discriminate between red and green colors and shades. In the case of red-green color blindness the green opsin gene is missing from the cell's genome or a green-red hybrid gene is in its place. (5)

In the mid-1980s geneticist Jeremy Nathans showed that the gene for red opsin resides on chromosome X right next to the gene for green opsin. (6) Individuals able to discern between red and green colors have at least one functioning copy of the red opsin gene and one functioning copy of the green opsin gene. Surprisingly, individuals with normal color eyesight can also have several copies of the green opsin gene following a single red opsin gene. In fact, three or more green opsin genes in a row are the norm.

Red-green color blindness and sickle-cell disorders are examples of only the simplest of genetic traits. Human genetics is generally far more complex; usually a multiplicity of genes interact with a complex environment. Genomics will help scientists identify much more rapidly the types of gene sequences like those that Jeremy Nathans unraveled to solve the puzzle of red-green color blindness. It is hoped that such knowledge will lead to cures and treatments for many disorders.

But genomics is also helping scientists understand complex functions including heart function, digestion, and growth. Driving each of these processes is a gene or series of genes that produces proteins that direct cellular and, eventually, organismal behavior. Before the genomic revolution scientists were limited by the technology of the time and could look at these complex interactions only one or, at best, a few genes at a time. Genomics is all about looking at hundreds, if not thousands, of genetic interactions simultaneously in order to understand the root causes of human disorders and to better understand how our bodies work.

Cancers illustrate the relationship between gene and genome. Most cancers are the result of several genetic mutations that can be caused by a combination of hereditary or age-related mutations and by environmental factors such as toxins, excessive sunlight, viruses, or diet. In most types of cancer, at least several genes must develop mutations for a malignancy to occur. One form of colon cancer, for example, generally begins with a benign tumor in the large intestine caused by a mutation on a gene known as APC, short for adenomatous polyposis coli. For these benign tumors to become malignant several other genes in the human genome must also have mutations. (7) Cancer may start with an error in one gene, but a series of genes interacting across the genome and with the environment is the reason that it can develop and proliferate. The human genome sequence and genomic technologies will help scientists better understand the exact genes involved in the development of human cancers. An understanding of the molecular basis of these cancers will also assist in developing drugs to treat cancers. We will discuss these methods in more detail in Chapter 7.

THE SEQUENCE

Despite phenomenal advances, gene sequencing efforts were still in their infancy in the mid-1980s. By 1988, just before the Human Genome Project got under way, laboratories owning even the most advanced technology could sequence only 50,000 nucleotides of DNA a year. (8) If this rate had held constant throughout the life of the project, it would have taken a single laboratory 64,000 years to sequence the entire complement of human DNA. Yet the technological advances of the previous decades inspired more and more scientists around the world to adopt and improve upon these technologies and take part in the rapidly growing field of molecular biology.

With the pieces of the gene sequencing puzzle falling into place and with new and faster technologies on the horizon, scientists began considering a coordinated effort to map the entire human genome. At a 1985 meeting hosted by biologist Robert Sinsheimer, then chancellor of the University of California, Santa Cruz, the idea of mapping and sequencing the human genome surfaced for the first time. (9) Sinsheimer, a distinguished biologist whose laboratory had mapped the genome of the bacterial virus Phi-X174, sought to create an institute to sequence the human genome on the Santa Cruz campus but was never able to stimulate the necessary interest within the University of California system. Sinsheimer's efforts did set something in motion, however, and "the idea of sequencing the human genome moved on to other pastures, having acquired a life of its own."

Other efforts to sequence the human genome, most notably a short-lived private venture led by Harvard biologist Walter Gilbert, continued to attract attention to the project. (10)

Beginning in the mid-1980s, through the visionary leadership of scientists like Sinsheimer, Gilbert, James Watson at Cold Spring Harbor Laboratories, Charles Delisi at the U.S. Department of Energy, and Renato Dulbecco at the Salk Institute, the idea of sequencing the human genome quickly gained adherents. In 1989, with legendary biologist James Watson at its helm, the National Institutes of Health created the National Center for Human Genome Research, which later received full institute status as the National Human Genome Research Institute (NHGRI). (11)

But the new genome institute and the growing emphasis on large-scale gene sequencing were not without detractors. Many social and natural scientists worried about the potential misuse of genetic information, whereas others worried that the project itself threatened the state of scientific research. Leslie Kozak from the Jackson Laboratory in Maine wrote that the project "threatened the quality and conduct of our nation's health-related research effort." Another critic wrote that the project was "mediocre science and terrible science policy." (12) Many biologists worried that such a large and centrally directed research effort would stifle the biological community, steering funds away from basic science and from important biomedical research. Sociologist Dorothy Nelkin was concerned about the effects of genetic technologies on privacy and discrimination, fearing that genomic research could help foster a "genetic underclass." (13) Even as James Watson declared that the genome would someday radically change medicine and science, many still worried that the genome's promise was far greater than its potential.

Despite the criticism, the Human Genome Project officially got under way on October 1, 1990, with a budget of almost $90 million per year. The bulk of the sequencing was divided between several facilities including those at NIH, the Whitehead Institute at MIT, Baylor College of Medicine in Houston, and Washington University in St. Louis. The project's objectives were based on a 1988 National Research Council report that listed three main goals of the Human Genome Project: (1) to construct a map and sequence of the human genome; (2) to develop technologies to "make the complete analysis of the human and other genomes feasible" and to use these technologies and discoveries to "make major contributions to many other areas of basic biology and biotechnology"; and (3) to focus on genetic approaches that compare human and nonhuman genomes, which are "essential for interpreting the information in the human genome." (14)

Scientists in the international community shared these goals as other large-scale genome projects developed in European nations, primarily in Britain at the Sanger Institute in Cambridge, and also in Japan. The participation of scientists from all nations, essential to the success of the project, was secured with the establishment of the Human Genome Organization (HUGO), founded in 1988, to

GenBank Data

Year	Number of bases deposited into GenBank, a public genome database
1982	680,338
1983	2,274,029
1984	3,368,765
1985	5,204,420
1986	9,615,371
1987	15,514,776
1988	23,800,000
1989	34,762,585
1990	49,179,285
1991	71,947,426
1992	101,008,486
1993	157,152,442
1994	217,102,462
1995	384,939,485
1996	651,972,984
1997	1,160,300,687
1998	2,008,761,784
1999	3,841,163,011
2000	11,101,066,288
2001	15,849,921,438
2002	28,507,990,166

This chart shows, year by year, the rapid pace in growth of gene sequencing.
(*Source:* http://www.ncbi.nlm.nih.gov/Genbank/genbankstats.html)

"promote international discussion and collaboration on scientific issues and topics crucial to the progress of the world-wide human genome initiative." (15)

MAKING THE GENOME POSSIBLE

The success of the Human Genome Project depended on new sequencing technologies. This was, in fact, one of the early goals of the project. The National Research Council's Committee on Mapping and Sequencing the Human Genome, whose program was adopted by the NIH, suggested that "a major portion of the initial monies should be devoted to improving technologies." The Committee also urged that "large-scale sequencing should be deferred until technical improvements make this effort appropriate." (16)

The then-current process of radioactive sequencing was too cumbersome and time consuming to be the primary method for sequencing the human genome. The sequences themselves were, after all, still being read by human eyes and recorded by hand. Leroy Hood's laboratory at the California Institute of Technology began to tackle this problem in the early 1980s and, through a series of significant technological improvements, established new methods to speed the pace of gene sequencing and eventually of the project itself. Hood's lab at Cal Tech contributed two critical tools to the sequencing revolution. First, the Hood lab developed a way to tag DNA nucleotides with fluorescent dyes instead of radioactivity. Thus a sequence's As, Ts, Gs, and Cs, could be tagged with four different fluorescent colors. This technology allowed the four colors to be read in one lane of a gel, allowing four times as many gene sequences to be generated. This greatly increased sequencing output. Second, they automated the process with computers. The four-color system uses a laser beam that scans over the fluorescently labeled DNA fragments. The laser beam "sees" reactions of different wavelengths based on the type of fluorescent dye and feeds these data into a computer. Simple computer programs translate the fluorescent wavelength data into the corresponding nucleotide sequence. (17)

In 1981, capitalizing on their success with automating protein sequencing, Hood formed Applied Biosystems. (18) After the development of the automated DNA sequencer Applied Biosystems and Hood's lab at Cal Tech worked together to perfect and streamline the device. In 1988 the first generation of automated gene sequencer, known as the ABI 377, went on the market. During the 1990s Mike Hunkapillar, once a scientist in the Hood lab and now head of Applied Biosystems, continued to work with his colleagues at Cal Tech to improve on the features of the 377. But the 377 could not handle the overwhelming quantity of sequences necessary to complete the human genome. For that task an entirely new technology was needed. Even before they designed the 377, Hood and Hunkapillar tried to develop a machine that used capillary tubes filled with polymer solution to run their sequences. In this process DNA fragments pass through tubes just a fraction wider than a human hair instead of through a gel. It took nearly two decades to develop a machine based on this technology: The ABI Prism 3700 increased sequencing capacity eightfold, allowing researchers to sequence as many as 1 million bases of DNA in 1 day. In 1998, at $300,000 per machine, the 3700 went on sale and immediately changed the face of genomics. (19)

RACING TO THE FINISH

The ABI Prism 3700 was the technological breakthrough everyone in the genomic community was hoping for. After the release of the new machine in 1998, an optimistic Francis Collins, who had succeeded James Watson as head of NHGRI, issued a five-year plan for the project, targeting 2001 for the completion of a draft sequence and 2003 for a full sequence of the human genome—two full years ahead

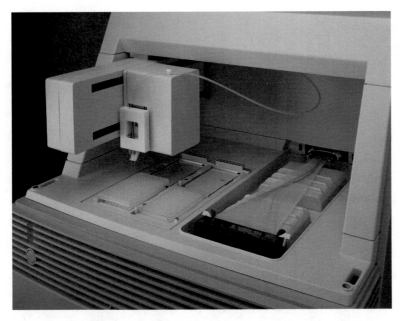

Hundreds of machines just like this one, the ABI 3700, were used to sequence the human genome.

of schedule. (20) The new machines quickly became the mainstay of genomic sequencing—several hundred 3700 machines could, after all, sequence the entire human genome in just a few years.

However; it was much more than this technological advance that accelerated the pace of sequencing. In May 1998 Craig Venter, a former NIH research scientist who had become president of The Institute for Genomic Research (TIGR), and Mike Hunkapillar of Applied Biosystems announced the formation of Celera Genomics. The new company planned to sequence the human genome privately by 2001—an effort both faster and cheaper than NHGRI's. (21) Such a claim quickly aroused serious concerns in the scientific community. Unlike NHGRI, which deposited its gene sequences every 24 hours into a public database known as GenBank, Celera would delay the release of its data for up to three months to allow paying customers, such as pharmaceutical and biotechnology companies, to access Celera's sequence data. Subscribers to Celera's database would get a head start in the search for genes that were good candidates for drug development. Celera also annotated its sequence, which according to a company executive, would make it "the definitive source of genomic and associated medical information." (22)

There were many researchers who considered the genome project to be an international effort, and all major sequencing centers around the world deposited their data daily into GenBank under the terms of the Bermuda Accord. Drawn up in 1996, this agreement states, "All human genome sequence information should

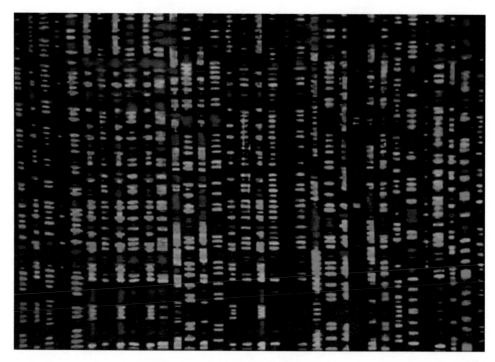

To the untrained eye, the vast code of our DNA is baffling. This rainbow array, the output from a gene sequencer, graphically represents the enormous volume of information in the human genome. Each color stands for one of the four bases of DNA—adenine, thymine, guanine, and cytosine, commonly abbreviated as A, T, G, and C. A laser in the ABI 3700 scans DNA samples tagged with four colors: A = green, T = red, G = yellow, and C = blue. A computer translates each color into the appropriate letter. Scientists then interpret and store the sequence in a computerized database.

be freely available and in the public domain in order to encourage research and development and to maximize its benefit to society." (23) If Celera was holding its data back to give paying customers an advantage, would that effect the development of new drugs and therapies? Would it stifle an open exchange of data? And what effect would the privatization of the genome have on the ethical challenges of genetics? University of Pennsylvania bioethicist Arthur Caplan asked, "Can the moral and legal questions [of the genome] be addressed if the largest scientific revolution of the next century is going to be done under private auspices?" (24)

Before cofounding Celera, Venter was well known in the molecular biology community. A Vietnam veteran and one-time avid surfer, Venter was a maverick both inside and outside the lab. (25) His tenure at NIH ended in 1992 after a dispute over a technique his lab had developed that detected sequences of regions of the genome that had been transcribed into RNA. Such regions are called expressed regions. The idea is that if you can locate these expressed sequence tags, or ESTs, you can find genes, because the majority of expressed molecules in cells are the result of the transcription of genes. Despite Venter's claim that he could locate 80–90 percent of expressed genes, critics believed the method to be of only limited use, one suggesting that the technique would be lucky to identify 10 percent of genes. (26) James Watson, then head of the Human Genome Project, testified before the Senate that the EST method "isn't science." Watson was concerned over the emphasis on automated sequencing, arguing that "virtually any monkey" could run the machines that locate ESTs. The tension between Venter and many in the genomics community was intensified by the decision to patent the ESTs. (27) Ironically, the NIH's Office of Technology Transfer was behind the patent application. The move to patent was, however, regarded as premature by many. Those who opposed the patents pointed out that the ESTs had no known function or utility and that this move would "undercut patent protection" and eventually "impede the open exchange of information on which the Human Genome Project depends." Geneticist David Botstein argued that "no one benefits from this, not science, not the biotech industry, not American competitiveness." (28)

That same year Watson rebuffed Venter again, denying him the right to use NIH grant money to do EST sequencing. With his research goals now limited by Watson and other government scientists, Venter left the NIH in 1992 and founded The Institute for Genomic Research (TIGR) with money from a venture capitalist. Venter's EST method quickly became a standard and widely used genomic technique. Even James Watson came to acknowledge his error, saying that the method "should have been encouraged." (29)

In his new position as president of TIGR, Venter turned his attention to sequencing an entire genome of an organism. In the hunt to find genes, the genome project was funding the sequencing of what are called model organisms, organisms with characteristics that facilitate genetic studies. Model organisms generally have small genomes and short generation times and can be bred quickly, characteristics that led genome scientists to believe that their DNA would be useful in helping locate genes in human DNA (Chapter 6 discusses this process in more detail). Mice are probably the most important model for biomedical research because of their close mammalian relationship to humans. Mice are, however, very expensive and laborious to work with. The zebrafish is useful to scientists because it is one of the few vertebrate model organisms and because its transparent embryo allows researchers to directly observe developmental processes and cellular interactions. The hermaphroditic nematode worm *C. elegans* allows for very fast and controlled breeding and is incredibly useful because of its cellular simplicity. Each individual always has the same number of somatic (nonreproductive) cells—959. (30) Each one of these cells can be identified and studied in detail. This simplicity allows scientists to more easily study the relationship between specific genes and development at the cellular and molecular levels.

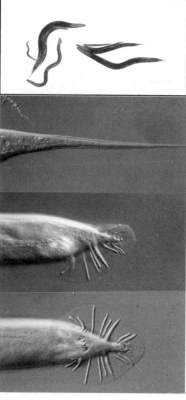

Studying model organisms, including the zebrafish, *C. elegans*, and *E. coli*, is helping scientists better understand gene function in all species including humans. The *C. elegans* pictured are each no more than 1 mm long. The *E. coli* image is from a model constructed for The Genome Revolution exhibit at the American Museum of Natural History. The gray capsulelike objects are a strain of *E. coli* shown in the lining of a human stomach.

By 1994, the model organism *E. coli* was in its ninth year of sequencing, a pace that Venter felt was too slow. That same year he and Nobel laureate Hamilton Smith applied to NIH for money to sequence *Haemophilus influenzae*, or *H. influenzae*, the cause of a deadly strain of bacterial meningitis in children. Venter and Smith claimed that they could sequence the *H. influenzae* genome in one year with a method known as shotgun sequencing. The NIH grant review board denied them funding, saying outright that their method and timetable would not work. Venter,

Smith and their team completed the genome sequence of *H. influenzae* in 1995, making it the first sequenced genome of a free-living organism. (31)

The greatest challenge to genome sequencing was finding the fastest and most accurate method for piecing together the data—over 3 billion bits of it in the human genome. The 3700s could supply data at an astonishing rate, but that information alone was just a meaningless string of As, Ts, Gs, and Cs. For the human genome to be of any use, strands of DNA needed to be organized by chromosome and by location on each chromosome. Two very different methods were used to accomplish this task. Until the successful sequencing of *H. influenzae*, many had considered the whole genome shotgun method an inferior and inexact sequencing technique. But in 1998 Venter and Celera boldly announced that they would sequence the human genome in just under three years using the whole genome shotgun method. At this rate Venter would sequence the genome several years ahead of the projected completion date of the NHGRI genome sequencing effort. (32)

The whole genome shotgun method relies both on the brute force of the ABI 3700 to quickly sequence billions of base pairs of human DNA and on complex computer algorithms that organize data rapidly and efficiently. The method has several steps:

Step 1: Unfragmented DNA from a single human genome was prepared. It was later revealed that Venter had sequenced his own DNA.

Step 2: The DNA was mechanically sheared into three sizes: 2000, 10,000, and 50,000 base pairs long.

Step 3: The different-sized DNA fragments were redundantly sequenced with ABI 3700s. The smallest-size fragments, at 2000 base pairs long, were primarily used to do the shotgun sequencing. The longer fragments were used as backups for filling in when gaps occurred. In some cases, the fragments were sequenced as many as 10 times to insure accuracy.

Step 4: Through computer algorithms, the genome was assembled, using the thousands of redundantly sequenced overlapping sequences. (33)

The NHGRI effort differed from the Celera effort in one major respect: Before mechanical shearing, 50,000-to 200,000-base pair-long sequences were laid onto chromosomal maps to determine genome location. After mapping, the fragments were broken down even further and then sequenced. In other words, segments of this reference map were sequenced piece by piece to discover and order all of the bases in the human genome. Celera's method did not use chromosomal maps. Instead, armed with previous knowledge of gene location and sequences throughout the human genome, Celera's algorithm was able to assemble the human genome without the physical mapping of Celera's source DNA. To fill in gaps in data, Human Genome Project scientists Robert Waterson, Eric Lander, and John

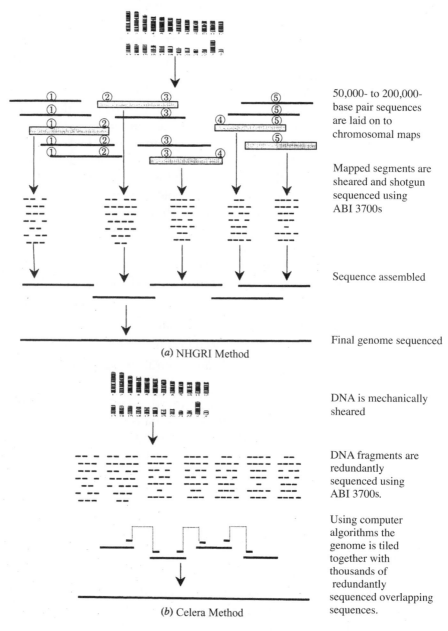

50,000- to 200,000-
base pair sequences
are laid on to
chromosomal maps

Mapped segments are
sheared and shotgun
sequenced using
ABI 3700s

Sequence assembled

Final genome sequenced

(*a*) NHGRI Method

DNA is mechanically
sheared

DNA fragments are
redundantly
sequenced using
ABI 3700s.

Using computer
algorithms the
genome is tiled
together with
thousands of
redundantly
sequenced overlapping
sequences.

(*b*) Celera Method

The NHGRI and Celera sequencing methods were different in one key respect:
Whereas the NHGRI genome effort used chromosomal maps to position the
sequenced DNA, Celera scientists used known genetic markers to assemble the
sequence.

Sulston have suggested that Celera used the public genome database to complete its draft sequence. (34)

The bulk of the NHGRI sequencing was completed at the Whitehead Institute for Biomedical Research at the Massachusetts Institute of Technology, the Sanger Institute at Cambridge University, and the Washington University Genome Sequencing Center. The Sanger Institute, for example, completed 30% of the genome, sequencing chromosomes 1, 6, 9, 10, 13, 20, 22, and X. (35) Seventeen other centers contributed to the sequence, including the Baylor College of Medicine Human Genome Sequencing Center, the Joint Genome Institute at the U.S. Department of Energy, the University of Washington Genome Center, the biotechnology company Genome Therapeutics, and laboratories in France, Germany, China, and Japan. (36) NHGRI coordinated the entire project.

In the months leading up to the completion of the draft, the media kept the public closely informed of developments. Headlines commonly focused on the race itself and its personalities: "New Company Joins Race to Sequence the Human Genome" (37) and "Competition Has Accelerated Race to Sequence the Human Genome." (38) William Haseltine, the president and CEO of the biotechnology company Human Genome Sciences, noted that "people may not understand genes or genomes, but they certainly understand a race." Craig Venter, at the height of competition between Celera and NHGRI, said, "They're trying to say it's not a race, right? But if two sailboats are sailing near each other, then by definition it's a race." (39)

Venter's rival in the genome race was geneticist-physician Francis Collins. Appointed NHGRI Director in 1993, Collins has overseen and coordinated the United States' most visible and important science project since the moon shot. He came to NHGRI as part of an elite club of molecular biologists who helped make genetics what it is today.

Collins is a gene hunter—he has spent his career locating genes that cause human diseases in the hope that his findings will eventually lead to cures. His distinguished work in this area includes codiscovery in 1989 of a gene responsible for cystic fibrosis. Involvement in the discovery of genes for Huntington disease, neurofibromatosis, and a form of adult leukemia followed. (40)

Collins and Venter shared identical goals—the sequencing of the human genome and the exploitation of genomic data for scientific and medical uses—but their paths to that goal had significant differences. At Celera, Venter delayed the release of sequence data for up to three months, giving his biotech and pharmaceutical subscribers time to mine the data for useful nuggets. Collins believed such a delay went against the spirit of science, and NHGRI data were put into GenBank, the public genome database, every 24 hours. (41) Differences aside, the competition turned out to be a boon to the genome project. Not only did it speed up the pace of sequencing, but media attention attracted the public to the story. Genomics suddenly became a household word as people around the world were reading about their genomes and considering the implications of the new technology.

In the end the race to sequence the genome was declared a draw. Over pizza and beer at the home of Ari Patrinos, the director of the Department of Energy's genome project, Venter and Collins decided to jointly announce the completion of the first draft sequence of the human genome. Although it came at the eleventh hour, Collins called the meeting "absolutely the right moment to sit down together." Venter agreed that it was "important for us to rise above the squabbles." (42) Although the competition to sequence the genome might have been a good thing for genomics, an ongoing public feud between these two scientific heavyweights would have tainted the project and distracted everyone from the task at hand. At a Rose Garden ceremony in June 2000, President Clinton celebrated the completion of the sequence itself, extolling the genome as "the language in which God created life" and acknowledging the important contributions made by both NHGRI and Celera. (43)

Both Venter and Collins continue to play leadership roles in genomics. Collins remains, as of 2004, head of the National Human Genome Research Institute, overseeing the ongoing Human Genome Project as well as other genomic initiatives. In 2002 Venter left Celera and formed several not-for-profit foundations that focus on different areas of genomic research. The Center for the Advancement of Genomics (TCAG) will continue Venter's work on sequencing genomes and developing other genomics projects. One of those projects is to sequence the genomes of 1000 individuals. TCAG also seeks to play an important role in exploring important social and public policy issues related to genomics. Venter's Institute for Biological Energy Alternatives is hoping to use genomic technology to develop clean energy sources. (44)

All this talk of the history of genetics, the race to finish the genome, and the ABCs of the gene may seem far removed from the ways in which the genomic revolution will affect our lives. We know what it took to sequence the genome, but what will all this effort mean? The greatest challenge now facing us is finding ways to benefit fully from advances in genomics and knowing when to be wary of potentially dangerous and disruptive technologies. To be sure, there is potential for great harm from genetic technologies. But the potential for harm should not limit our vision of the future or supersede a belief that genomics can and will benefit humanity greatly. The difficult choices ahead may, for example, involve limiting or forestalling the development of particular technologies. How will government, business, or the public decide how to do this? Will genomics necessitate new standards of safety for potentially dangerous technologies? How will individuals and groups of people be kept safe from genetic discrimination? These are just some of the questions that policy makers, scientists, clergy, and government officials are beginning to consider as discoveries in genomics begin to affect all of our lives.

PART 2

Information

We are living in an age of information. From the ease with which the Internet makes available the most obscure or personal facts to the abundance of media outlets, technology permits and maybe even encourages instantaneous and sometimes gratuitous access to information. We are only beginning to understand, however, the ways in which this technology is transforming our selves and our society. How is this ease of access affecting our understanding of the world around us? And is the world becoming a smaller place because of the information revolution, or is it becoming a place that isolates us as we spend more time sitting in front of computers and televisions?

Like the flood of digital and media information, the torrent of data being generated by genomics will have a profound effect on our lives. The information contained in our genomes and in the genomes of other species will present many social challenges. Access to genetic information not only holds promise for great advances in medicine and science but has great potential for misuse. Genetic discrimination in health insurance and employment may someday become a reality. Genetic engineering introduces profound moral, theological, and ecological dilemmas. And, finally, our genomes themselves are changing the ways we think about ourselves and our relationship to all the other species on earth. The three chapters in this section explore these issues. Chapter 4 looks at important ethical and moral issues generated by genomic data and the ways society is preparing to grapple with and manage this knowledge. Chapter 5 considers the ways in which science and society think about the relationships between and among human population groups, examining specifically the idea of race in a genomic world. Chapter 6 looks at the way knowledge of the genome is changing how humans think about themselves in relationship to all the other species on earth.

4

Keeping the Genome Safe

A report issued by the National Science Foundation in 2001 found that 89 percent of Americans believe that "scientists want to make life better for the average person." The report also indicated that 72 percent of Americans have faith that the "benefits of science are greater than the harmful effects." (1) These sentiments, we suspect, are based on the continued progress of scientific research as well as a genuine optimism that science can and will help treat or cure the diseases that affect all of our lives in some way. Americans express that continued faith with overwhelming support of federal funding for scientific research and a keen trust in research scientists and medical doctors. (2)

The success of the genomic revolution depends, in large part, on the general public's continued faith in science. As subjects in clinical research, as voters with a say in public spending, as interested observers, and as consumers of health care we are all participants in the genomic revolution. But what would happen if the public no longer believed that science is making life better for the average person and that the benefits of science outweigh its harmful effects?

As great as the benefits of genomic technology may be, so too are its possible misuses. Eugenics, genetic discrimination, and other abuses of genetic privacy have the potential to do great harm. Genomics will also force us to consider the limits of applying our knowledge. Where should the line be drawn between acceptable and unacceptable applications of genomics, and who should draw it? Should we, for example, continue to develop technologies that allow us to screen for traits in embryos? Should we, if possible, manipulate those traits? And if the technology is feasible, should we develop genetic enhancements that do not simply treat disease but make us "better than well," enhancing humanity in ways foreseen only in science fiction? (3) In this chapter we will highlight some of the moral, ethical, and policy challenges of genomics and look at how our society is preparing itself to answer such questions and to ensure that science and medicine continue to work to improve human health and minimize human harm.

61

Welcome to the Genome, by Rob DeSalle and Michael Yudell.
ISBN: 0-471-45331-5 Copyright © 2005 Rob DeSalle and Michael Yudell.

A SHAMEFUL PAST

Beginning in late 1946, 23 Nazis were put on trial for participation in human experiments that killed thousands of Jews, Gypsies (now more commonly referred to as Romany), homosexuals, and mentally and physically disabled people, among others. (4) The research, carried out both to support the German war machine and to satisfy the malevolent curiosities of German scientists, was intimately connected to the German mind-set of the time. In the madness of Nazi Germany, the integration of eugenic ideas into popular thought meant that the so-called unfit were exploitable and, ultimately, expendable. A national culture of hatred toward Jews and others identified by Hitler as a threat to the German people shaped the behavior of scientific professionals in ways that we hope we never see again. (5)

Concentration camp prisoners were forced to participate in horrific experiments. To study the effects of high altitude on German pilots, unknowing subjects were placed in decompression chambers. Some victims were killed by evacuating the air from the chamber, others by having their lungs dissected while they were still alive to study the chamber's effects. (6) In another experiment conducted at the Dachau concentration camp, inmates were used to study the effects of hypothermia. Men were placed in a vat of ice and water and monitored until they lost consciousness. Researchers then tested different techniques to revive those who were still alive. The dead were autopsied. (7)

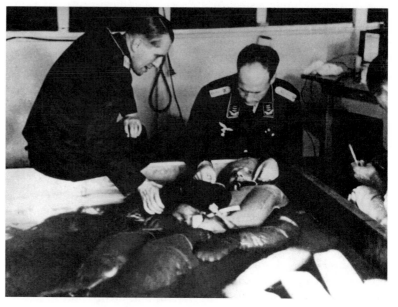

At the Dachau concentration camp prisoners were used in brutal experiments, against their will, to study the effects of hypothermia.

At the Auschwitz-Birkenau concentration camp, the infamous Dr. Joseph Mengele conducted experiments on twins "intended to demonstrate a hereditary basis for group differences in behavioral and physical characteristics." (8) To that end, Mengele compared medical outcomes on twin pairs, exposing one or the other to X-rays and deadly diseases. In one particularly cruel episode, Mengele stitched together, back-to-back, two opposite-sex twins. (9)

These, just some of the horrors that came to light during the Nazi doctors' trials, stand as testament to the suffering of victims of these atrocities. But out of the trials came something that would also honor the memories of those abused—a comprehensive code of conduct for research with human subjects. The Nuremberg Code, written by the Nuremberg trial judges, proposed 10 directives that still shape research with human subjects. At the core of the Code is the principle that "voluntary consent of the human subject is absolutely essential" in a research protocol. The Code also requires that an experiment be "for the good of society" and "not random and unnecessary in nature," that the experiment "avoid all unnecessary physical and mental suffering and injury," that subjects be protected "against even remote possibilities of injury, disability, and death," and that at any time during the experiment the subject can discontinue his or her participation in the study. (10)

Unfortunately, the Third Reich was not alone in its mistreatment of research subjects. Since World War II research subjects in many nations, including the United States, have been ill-treated because of lapses in judgment and ethics. In the United States, for example, science and medicine could and sometimes did engage in unethical and immoral behavior, but such experiments were not a systematic and integrated part of the scientific and medical establishment. Despite the best intentions of the Nuremberg Code, the behavior of many scientific and medical researchers suggests that the short-term impact of the Code was limited. Two events stand out for their ethical lapses that ultimately led to the establishment of lasting research standards. In 1966 the *New England Journal of Medicine* published a shocking exposé entitled "Ethics and Medical Research" by the distinguished anesthesiologist Henry Beecher, who held positions at Harvard Medical School and Massachusetts General Hospital. Beecher detailed 22 cases of egregious ethical lapses in research on human subjects that had taken place over the previous two decades at prestigious medical schools, university hospitals, and the National Institutes of Health. (11)

In the years following World War II, American scientific and medical research flourished, spurred on by significant increases in government spending. (12) This change brought with it both "vast opportunities and concomitantly expanded responsibilities," which Beecher worried were not always being met. In one example, Beecher cited a case where institutionalized mentally retarded children were purposely infected with hepatitis to determine the disease's "period of infectivity." Even though the parents of the children gave consent for the injection of the virus, they were not told of the hazards involved. In another study, 22 patients were injected with live cancer cells "as part of a study of immunity to cancer." The patients were not told that the cells were cancerous. Finally, in a curiosity study which had no known medical benefit, 26 newborns less than 48 hours old with normal bladders were catheterized and X-rayed to see whether a particular urinary condition can occur even in normal children. (13)

Just six years later a more shocking revelation came to light. For 40 years beginning in 1932, the United States Public Health Service followed the progression of end-stage syphilis in 399 black men in Macon County, Alabama, to study the complications associated with the disease's final stage. No treatment was provided to any of the study subjects, even after penicillin was found to be effective against the disease in the 1940s. Participants in the study were generally poor, illiterate African Americans who had been offered incentives by the Public Health Service to participate, including money paid to their survivors for burial costs. Many of the men were never even told they suffered from syphilis. (14) Known as the Tuskegee Study, this research was never secret, and its findings were published periodically in the nation's leading medical journals. Historian James Jones points out that "not since the Nuremberg trials of Nazi scientists had the American people been confronted with a medical cause célèbre that captured so many headlines and sparked so much discussion. For many it was a shocking revelation of the potential for scientific abuse in their own country." (15)

The cases revealed by Beecher and the Tuskegee Study are examples of research at its worst—investigators sacrificing the safety of their subjects for what in their judgment was a greater good. Columbia University historian and bioethicist David Rothman points out that much has changed since these events were revealed and suggests that "the experiments that Henry Beecher described could not now occur; even the most ambitious or confident investigator would not today put forward such protocols." (16) If so, the reason is that Beecher's exposé and the revelation of the Tuskegee Study helped provoke lasting institutional and cultural changes in scientific and medical research, including federal policies developed in the 1970s that established regulations for protecting human research subjects. (17) These new policies grew out of the National Commission for the Protection of Human Subjects and are administered by the Department of Health and Human Services. They include reliance on institutional review boards, or IRBs, that screen research proposals and monitor ongoing studies to ensure ethical treatment of human subjects and a series of ethical principles to govern all human research protocols. (18)

As articulated in the National Commission's 1979 Belmont Report, three governing principles—respect for persons, beneficence, and justice—have become the foundation of biomedical ethics. Respect for persons means that people should be treated as individuals able to determine what is best for them and that people with "diminished autonomy," such as the mentally ill, deserve special protections.

Beneficence means that a study should do no harm and should "maximize benefits and minimize possible harms." Justice refers to the fair distribution of the benefits and burdens of research. This means that, for example, "the selection of research subjects needs to be scrutinized in order to determine whether some classes (e.g., welfare patients, particular racial and ethnic minorities, or persons confined to institutions) are being systematically selected simply because of their easy availability, their compromised position, or their manipulability, rather than for reasons directly related to the problem being studied." (19)

When the Human Genome Project got under way, many feared that the abuses highlighted by Beecher or those that took place during the Tuskegee Study would pale in comparison to what a genetic revolution would bring. One author warned that the genome project raises "the likelihood of a new form of discrimination." (20) Another feared that "eugenics will come to America as a homespun mom and pop operation to weed out the less than perfect." (21) The stakes were and continue to be enormous. In the early 1960s the distinguished evolutionary biologist Julian Huxley recognized that through genetics humankind could come to control its own future, calling our species "the trustee... of advance in the cosmic process of evolution." (22) Such dangers and difficult choices would necessitate not only a coordinated international effort by scientists but also the participation of bioethicists, theologians, and public policy makers in helping to understand the nature of the dangers involved in genomic research.

PREPARING FOR THE GENOME

At a 1988 press conference James Watson, who had just been appointed head of the Human Genome Project, made a remarkable announcement: At least 3% of the project's annual budget would go toward the study of the ethical challenges genomics might raise and the formulation of policies to address these challenges. (23) Formally established in 1990, the Ethical, Legal, and Social Implications Program of the Human Genome Project, or ELSI for short, immediately became the largest bioethics program ever. ELSI works primarily as a genome project program that funds studies by academics and policy makers. According to bioethicist Eric Juengst, the purpose of the ELSI Program is "to anticipate and address" the "implications of acquiring and using" genetic knowledge "to help optimize the benefits to human welfare and opportunity from the new knowledge, and to guard against its misuses." (24)

ELSI research has focused primarily on four areas: (1) genetic privacy and fair use of genetic information in health care and employment, (2) the integration of genetic technologies into clinical care, (3) the protection of human subjects in genetic research, and (4) the promotion of public and professional education about genetics. (25) Since its inception the ELSI Program at NHGRI has distributed more than $100 million in research grants to examine these topics and to develop related books and literature. (26) It has also funded conferences, task forces, films, and research articles and has had a direct effect on the development of policies and laws relating to genetics and the human genome. Scholars supported by ELSI, for example, developed a prototype Genetic Privacy Act that was the model used by many states in adopting anti-genetic discrimination laws. (27) As of early 2004, 32 states had some form of anti-genetic discrimination in employment law and 34 states had some form of anti-genetic discrimination in health insurance law. (28) Although all of these laws prohibit discrimination against workers based on the results of a genetic test, the details of these laws vary from state to state. Most of the state laws also restrict employers from access to genetic information.

ELSI-funded scholars and others have paid careful attention to the nature of research ethics in the genomic age. United States federal regulations require that participants in scientific and medical research give their informed consent by affirmatively acknowledging that they understand the nature of the research and are told the risks and benefits of participation. (29) Many experts, including bioethicists and legal scholars, are concerned that genetic research with human subjects may require additional measures to protect those both directly and indirectly involved in such research.

Unlike most basic research, genetic research can involve risks that continue past the completion of a study. (30) For example, genetic data may reveal information that has an impact on family relationships or on an individual's medical future. A man whose father had died from a confirmed case of Huntington disease was genetically tested for the illness. The test revealed that the man was not at risk for the disease. The test also showed, however, that the man was not biologically related to his father. (31) The consent form the man signed did not reveal this possibility, so the man was not informed of this particular result. What if the testing also revealed the man was at risk for another illness? Should the researchers share this information with the tested individual? Consent forms are currently not consistent on this issue, and "researchers are generally given the choice of telling their research subjects... that they will, will not, or might if they choose to return information to them." (32)

People may also want to know how their genetic samples are being used. Even if confidentiality of the samples is maintained, some people may not want their genes being studied in certain ways. For example, a person who participates in a genetics of heart disease study may or may not want his or her genes used to examine the genetics of drug addiction, intelligence, or sexuality. (33) An informed consent document will generally notify study participants if their genetic material can be used at a later date for other studies. But with the advent of large DNA data banks designed as a resource for scientists to study multiple genetic effects, the nature of consent is different. In such cases individuals who provide genetic samples and medical histories may do so by giving blanket consent, allowing the DNA bank to share the same sample in different types of genetic studies. Critics warn, however, that blanket consent does "not allow patients to act meaningfully on their continuing right to control their health information." (34) Such concerns are being considered in the development of private genomic databases. For example, First Genetic Trust, a private DNA data bank that supplies genetic samples and related health information to researchers, allows contributors to restrict the use of their DNA sample depending on the nature of the research. (35)

Informed consent is also facing scrutiny as it pertains to genetic research that presents risks to more than just the individual participating in a study. Studies that examine the relationship between a particular racial or population group and a disease may have the unwanted side effect of stigmatizing that group. Some bioethicists have suggested that, when possible, group consent or community review complement informed consent. (36)

Despite the tremendous success of ELSI as a resource helping to identify the legal, social, and ethical challenges of genomics, some have voiced concern that the program is a "watchdog that's all bark and no bite." Others worry that the ELSI budget was taking funds away from scientific projects. (37) Yet ELSI has provided both scholars and policy makers with an unprecedented resource for developing regulations and guidelines. ELSI also serves as an important model for demonstrating that the integration of ethics and policy initiatives into science can be a practical and valuable component of scientific and medical research.

Ethics and policy programs have also been established in other countries. The Wellcome Trust Biomedical Ethics Program was founded in 1997 in the United Kingdom to support studies on ethical, legal, social, and public-policy issues of biomedical science, including genetics. Other countries, including Canada, France, and Germany, have some form of government funded genome policy and ethics program. (38)

There remains a concern that the technology is moving ahead at such a fast pace that we cannot guarantee its safety and may not be ethically or morally comfortable with where it leads us. With so many different stakeholders in the genomic revolution, it is unlikely that we could arrest the development of a particular technology even if we wanted to. We can, however, do our best to guide new technologies in a way that prevents us from formulating policies that are reactions to some type of catastrophe brought about by an accident or by misuse of genomic technology. There is an important precedent in the recent history of science along such lines.

In 1973 biologists Herbert Boyer of the University of California and Stanley Cohen of Stanford University were the first to successfully genetically engineer an organism by moving a gene from one bacteria into another. (39) At the time, it was recognized that this advance held great promise for both medicine and agriculture. But Boyer and Cohen were among those who worried that this type of genetic engineering, or recombinant DNA technology as it is also known, posed great safety concerns. As a result, leading molecular biologists, working with the National Academy of Sciences, called for a voluntary moratorium on the development of genetically engineered organisms that could introduce new antibiotic resistance or new bacterial toxins into nature or of genetically engineered organisms that could introduce cancer-causing or other animal viruses into the DNA of any organism. (40)

While the moratorium continued, scientists, policy makers, and others gathered in 1975 at the Asilomar Conference Center in Pacific Grove, California, to investigate the "potential dangers" of genetic engineering and to begin to develop regulatory guidelines that would allow the research to go forward safely.

A group of distinguished scientists and policy makers, including James Watson (left) and Sydney Brenner (right), gathered at the Asilomar conference center in Pacific Grove, California in 1975 to discuss biotechnology safety. The group recommended a temporary halt to certain types of recombinant DNA research until its safety could be ensured.

As a result of the Asilomar conference, the National Institutes of Health Recombinant DNA Advisory Committee, or NIH-RAC, which had been formed in 1974, was given the responsibility to develop safety guidelines for all recombinant DNA work taking place at institutions receiving NIH funding. (41) When the NIH-RAC guidelines were published in 1976, the moratorium on recombinant DNA research was lifted. (42)

The guidelines developed at Asilomar have had a lasting impact. Recombinant DNA technology remains a safe and reliable technology. The guidelines also illustrate the important role that scientists themselves can have in identifying and helping to regulate technologies that may be a threat to the public health. During the genomic revolution, scientists and policy makers may together decide to declare a similar moratorium on technologies like cloning or some types of genetic engineering because of safety, ethical, or moral concerns. Despite our great trust in science and medicine, we know from history that technological advance is often fraught with great danger. In the following examples we outline some of the potential hazards of genomics and discuss the ways in which we might be able to prevent history from repeating itself.

GENETIC DISCRIMINATION

In February 2001 the U.S. Equal Employment Opportunity Commission, or EEOC, sued the Burlington Northern and Santa Fe Railway Company for violating the Americans with Disabilities Act by infringing the genetic privacy of its workers. At issue was the testing of workers diagnosed with the repetitive hand motion injury carpal tunnel syndrome. Through secret genetic testing, the company hoped to show that the syndrome was a preexisting condition, not an on-the-job injury, thereby excusing itself from paying for its workers' medical treatments or disability claims. Among the details of the settlement reached in May 2002 was a requirement that the company discontinue its genetic testing program and that it pay $2.3 million to the workers who had been tested. (43)

The case brought to light what many feared would be one of the downsides of the genomic revolution: Without an individual's consent genetic tests could be conducted and used to discriminate against that individual. Although in the end the EEOC's intervention prevented genetic discrimination at Burlington Northern, the collection of the workers' DNA through a routine blood draw shows just how easy it is to conduct such tests. Because genetic tests have the capacity to both diagnose preexisting conditions and forecast future health, there is, as the Burlington Northern case shows, great potential for misuse. Burlington Northern's management hoped that a genetic test could show that a person's biology is more significant than his or her working conditions when it comes to this particular disorder. In other cases, genetic testing could be used to predict an individual's future health. Potentially such information could then be used either to deny health insurance based on a preexisting condition or to make employment decisions adverse to the worker.

We mentioned above that many states have already made great strides in prohibiting various forms of genetic discrimination. However, there are still no comprehensive federal anti-genetic discrimination laws in employment and health insurance. Nevertheless, legislators in Washington have been working toward developing policy in this area. In 1995, for example, the EEOC issued an interpretation of the Americans with Disabilities Act of 1990 (ADA) that would outlaw employment discrimination on the basis of a person's genes. The ADA protects carriers of genetic predispositions who can show that they were "regarded as disabled" because of that predisposition. (44) The Health Insurance Portability and Accountability Act of 1996 (HIPAA), which applies only to employer-sponsored group health insurance plans, remains the only federal law that prohibits genetic discrimination by insurance companies. HIPAA protects individuals from being denied insurance based on the result of a genetic test. (45) Finally, President Clinton issued an Executive Order on February 8, 2000, prohibiting the use of genetic information in decisions regarding the hiring or promotion of all federal workers. (46) In October 2003, the United States Senate passed a bill that would prohibit employers from using genetic tests in hiring and other employment practices or using such tests to determine eligibility or status of health insurance coverage. The bill is now pending in the House of Representatives. (47)

Studies suggest that most people do not want insurers or employers to have access to genetic information because they fear discrimination. (48) Kathy Hudson, Director of the Genetics and Public Policy Center at Johns Hopkins University, worries that patients' fear of the misuse of genetic information "has weighed heavily in their [patients'] decisions about whether to have genetic tests that could improve their health." Fears like this could have a ripple effect, prompting individuals to lose confidence in genetic technologies, which in turn could affect the development and integration of genomics into medicine. Fear could also stop people from participating in important genetics research or from sharing pertinent genetic information with their families and doctors. (49) Federal anti-genetic discrimination laws, says Francis Collins, will remove this barrier and "tell people that it's safe to know about your own genome." (50)

Genetic discrimination can also have an impact outside employment and insurance decisions. For example, 12 states and the District of Columbia mandated screening for sickle-cell anemia for African Americans in the early 1970s. (51) These laws singled out African Americans despite the fact that variants of sickle-cell and other blood anemias have a significant effect on other population groups including Greeks, Indians, Arabs, and Italians. These laws were eventually repealed, but they show how genetic information can be used to stigmatize certain groups. In another case, the Lawrence Berkeley Laboratories in Berkeley, California, an important center for genomic research, was found in 1999 to have violated Title VII of the Civil Rights Act of 1964 for singling out its African American employees for a genetic test for sickle-cell anemia. (52) These types of testing programs reinforce "inappropriate stereotypes about African Americans being genetically inferior." (53)

THE BUSINESS OF BIOLOGY

On September 17, 1999, Jesse Gelsinger, a 19-year-old patient in an experimental gene therapy protocol at the University of Pennsylvania's Institute for Human Gene Therapy, died from a severe immune reaction within hours of undergoing treatment. Gelsinger had been born with the rare metabolic disorder ornithine transcarbamoylase (OTC) deficiency. Babies born without the capacity to produce the metabolic enzyme OTC usually die soon after birth, but patients like Gelsinger, who are only deficient in the enzyme, can survive on a strict diet with drug treatment.

Nineteen-year-old Jesse Gelsinger was the first patient to die as the result of a gene therapy experiment. Gelsinger's death provoked controversy that highlighted the conflicts of interest that can threaten the safety and integrity of biomedical research.

(54) Although Gelsinger did not need the gene therapy to survive, he volunteered for a Phase 1 safety trial, hoping that his participation might someday improve the quality of his life and help save the lives of those born with the more severe form of OTC. (55)

Gelsinger's was the first known death from a gene therapy experiment, and it raised alarms throughout the field. Despite important successes in gene therapy research since Gelsinger's death (see Chapter 7), Gelsinger's participation in the OTC trials drew attention to the relationship between research and profit, between biology and business. Beginning with the growth of the biotechnology sector in the late 1970s, and more recently with the rise of genomic technology, the line between academic and private science has blurred. When research scientists stand to benefit financially from the outcome of their research, many worry that profits will trump safety and ethics. That is what appears to have happened in the tragic case of Jesse Gelsinger.

An investigation conducted by the Food and Drug Administration revealed that the research team at the University of Pennsylvania violated several federal research rules. Four patients who had received the treatment before Gelsinger had had reactions to the treatment so severe that the trial should have been immediately stopped. The informed consent forms that should have told Gelsinger and others about all the risks associated with the trial were edited without FDA knowledge to eliminate mention of primates that had died because of receiving a similar treatment. And Gelsinger was included in the study despite the fact that his blood ammonia levels were above those considered safe for his participation. (56) Moreover, it was also revealed that James Wilson, a principal investigator of the study, had a major financial interest in the outcome of the trial both as head of Penn's Institute for Human Gene Therapy and as a founder of Genovo, a biotechnology company that held patents on the OTC treatment. (57)

In November 2000, the Gelsinger family settled a lawsuit against the University of Pennsylvania, Genovo, Wilson, and some of the other researchers involved in the study. (58) The scientific community's reaction to the Gelsinger case was also far-reaching. Policies to control conflicts of interest are now more commonplace. The Association of American Medical Colleges, for example, issued a statement in December 2001 calling for universities to disqualify a researcher's involvement in a clinical trial if the researcher held any interest in companies with a stake in the trial's results. Many major medical journals now require authors to disclose any possible financial conflicts of interest. The U.S. Department of Health and Human Services is developing policy guidelines for financial conflicts of interest in human subjects research. (59) As the biotechnology and pharmaceutical industries capitalize on discoveries generated by the genome boom, this type of oversight will be crucial. Only by eliminating circumstances in which financial incentives outweigh or compete with patient safety can researchers help ensure that there will be no more tragedies like that which befell Jesse Gelsinger.

Gene patenting is another important "business of biology" issue. Many people have an almost visceral reaction against the patenting of human genes, as if it somehow violates our humanity to turn our genetic code into a business arrangement. Others worry that patents might grant large corporations control over our common genetic heritage. As a protest against patent policy, an antipatent protestor in Great Britain filed a patent claim on herself in the hope of maintaining "sole control" over her own genetic material. (60) As a legal matter, however, in the United States and in most other countries, the issue of patenting genes is, for the moment, settled.

Genes cannot be patented as they exist in nature. The United States prohibits the patenting of "products of nature." Thus gene patents are issued on what are essentially copies of genes, not the actual genes themselves. You could not, therefore, patent your own genome as it exists in your cells. You would have to, in essence, sequence your own DNA and identify particular genes to begin such a process. (61)

On the face of it, gene patents would seem to be a good thing. Ideally, patent protection would "provide incentives for innovation and for the development of products from which the public can benefit." (62) In the case of gene patents, it is hoped that investments by biotech companies and others over the 20 years of exclusionary patent rights will eventually lead to new diagnostic tests, therapies, or products. For example, Genentech's development of recombinant insulin, for which it had the exclusive patent rights, has far-reaching medical benefits.

Some critics point out, however, that it may not always work this way. New York University sociologist Dorothy Nelkin wrote that "a researcher who owns a patent on a gene or DNA sequence can prohibit others from using the gene or can charge high licensing fees to researchers who later try to develop related tests or therapies." (63) In such cases, gene patents could prevent the development of life-saving technologies. Gene patents might also stifle "life-saving innovations. . . in the course of research and product development." This could happen, for example, if the permission of several patent holders is required to create a particular gene therapy. If one of the patent holders withheld consent, then the therapy could not be produced under current patent rules, "potentially harming both the patent holder and the patent users." (64)

If it becomes clear at some point in the future that gene patents are inhibiting technological development and thus threaten the public health, Congress or the courts will have to intervene and develop some type of remedy. The U.S. Patent and Trademark Office recently made it more difficult to receive a gene patent, requiring applicants, in most cases, to show that their gene sequence has a utility that may be of some biomedical benefit. (65) This put a stop to the practice of patenting identified genes whose function was not always known. As the technology and our understanding of genomics evolves, so too may our legal requirements for patenting genes.

NATURE/NURTURE

We sometimes hear people say things like "Creativity runs in the family," or "I inherited this bad back," or "All the women in my family live past 80." People often wonder how they acquire their traits, from talents to ailments. The genes we inherit from our parents do indeed guide how our bodies develop and function, but where we live, what we do, and our individual environment, starting in the womb, also play a large role in determining our makeup. Nutrition, exercise, and education are just some of the influences on our health and behavior. Identical twins, for example, have the same genes, but twins develop unique personalities, disabilities, skills, and sometimes looks because of environmental factors. (66)

Researchers are finding connections between genes and human characteristics ranging from aging to drug addiction to disease susceptibility. For the most part, though, our genes are not our ultimate fate. We are instead a product of interactions between genes *and* our environment, nature *and* nurture. Harvard University evolutionary biologist Richard Lewontin sees this interaction as a triple helix, suggesting that an organism is the product of a "unique interaction between the genes it carries," the "external environments through which it passes during its life," and the random "molecular interactions within individual cells." (67) If we understand our individuality in this way, it is impossible to describe genes as the sole arbiters of our fate. Although genes can and do affect our lives, they are only a part of what makes us who we are.

One of the dangers of the genomic revolution is that people may place unwarranted faith in the power of genetics to heal and explain long-debated ideas about human nature. To be sure, genomics will cure diseases and will help untangle our understanding of who we are, but the lure of the all-powerful, all-explanatory gene is ultimately misleading. And this type of faith in genetics has had consequences. In the United States during the first third of the twenteeth century, the eugenics movement promulgated hereditarian theories of human social order and behavior in order to further its political and social agenda. And in Nazi Germany the "eugenics movement prompted the sterilization of several hundred thousand people and helped lead to anti-Semitic programs of euthanasia and ultimately, of course, death camps." (68)

Genetic determinism does not have to be so extreme to have damaging consequences. In 1969, Arthur Jensen, an educational psychologist at the University of California, maintained that racial differences in intelligence, or IQ, were hereditary. Educational efforts to raise IQ scores were, therefore, useless. (69) Jensen's theories have, over the past three decades, played an influential role in shaping the debates about educational priorities despite the clear evidence that shows that race and intelligence are not linked by genetics. (70)

The theories of sociobiology and evolutionary psychology provide a modern framework for genetic determinism. Sociobiologists and evolutionary psychologists claim that "the most diagnostic features of human behavior evolved by natural selection and are today constrained throughout the species by particular sets of genes." (71) However, like the theories of eugenicists, the genetic determinism of sociobiologists and evolutionary psychologists fails to account for what most natural and social scientists have recognized for decades: Complex social behaviors in humans, like aggression, sexuality, and ethics, are best understood as culturally and historically contingent rather than as discrete biological phenomena. Furthermore, the genome project is demonstrating that most complex traits and diseases cannot be accounted for in a single-gene, single-trait fashion and is spurring on a shift in the language of genetic causation. Whereas many scientists used to talk about the gene for a particular trait or disease, today they speak instead of genetic components.

Perfect pitch and drug addiction illustrate the role genes do play in complex human behaviors. It's tempting to think there must be a genetic component to account for a musical prodigy's extraordinary ability, but so far scientists can't point to any genes involved in musical talent. One exception seems to be perfect pitch: Research shows that the ability to "identify the pitch of a note instantly, accurately, and without the help of a reference tone" is inherited. (72) Jazz singer Ella Fitzgerald had perfect pitch, found in only 1 in 10,000 Americans. But nature aside, all aspiring musicians need training and practice, and individuals with perfect pitch require early training to develop this rare talent. Children are most likely to develop perfect pitch when they begin musical training by age six. (73)

Drug addiction offers another example of the role genes play in human behavior and of the nexus between nature and nurture. Studies have shown that worldwide between 1 in 3 and 1 in 5 people who try heroin become addicted to it. (74) Evidence suggests that this addiction is at the same time an environmental disease, a disease with a significant genetic component, and a disease that has physiological manifestations on brain structure. Each of these components of heroin addiction plays an important role in the course of the disease. The environmental influence, which can be some form of stress experienced at any point in life, can be a trigger that leads an individual to self-administer heroin. The genetic influence seems to be related to the way in which heroin interacts with the mu opioid receptor, a regulatory receptor on the surface of brain cells that plays a role in producing the euphoric effect of the drug. Recent studies have shown that genetic variation in the mu opioid receptor gene can have an influence on an individual's reaction to heroin. (75) Finally, studies have shown that heroin use can permanently alter molecular physiology, biochemistry, neurochemistry, cellular physiology, and behavioral physiology. (76)

Until recently, drug addiction was considered to be either deviant behavior or a personality disorder. Research over the past few decades has shown, however, that "addictions are disease of the brain." (77) This does not necessarily mean, however, that an individual who self-administers a narcotic and then develops an addiction had originally sought out the drug because of some underlying genetic need. An individual must first self-administer a drug and then have some kind of susceptibility, environmental or genetic, to become an addict. This complex web of environment, genes, and organism demonstrates just how hard it is to emphasize one component over the others in their impact on the disease. Together these components represent important avenues in the ongoing efforts by researchers to treat the disease of heroin addiction.

EUGENICS IN THE GENOMIC AGE

What most frightens observers of the genomic revolution is the prospect of genetically engineering human traits and redirecting human evolution through changes in our genes. With such tools humanity may come to control the very nature of our being, challenging notions of what it means to be human and perhaps someday being bystanders at our own recreation. If the technology to do such things can be developed, what will stop us from doing them? Princeton University molecular biologist Lee Silver proposes that in the future genetic engineering will divide humanity into two new species: the gene enriched, or GenRich, who are genetically enhanced with traits such as increased intelligence, resistance to disease, and superior athletic ability, and the Naturals, those consigned by class to their nonengineered "normal" fates. (78) Although this scenario may seem like science fiction, Silver suggests that the use of these "technologies is inevitable" and that it cannot be "controlled by governments or societies or even the scientists who create it." (79) After all, who would not want the very best that science and medicine can offer for both themselves and for their children?

But missing in this version of an inescapable future is the idea that we are at all actors in this unfolding science fiction drama. Science and technology do not develop on their own: They are products of human imagination and ingenuity. We should not assume that they cannot be subdued or stopped if we deem them dangerous in some way. Bioethicist Leon Kass worries that the "introduction of new technologies often appears to be the result of no decision whatsoever, or the culmination of decisions too small or unconscious to be recognized as such. Fate seems to hold the reins." (80) If we should allow some genomic technologies to take hold in this way, humanity may someday regret its acquiescence to a technological imperative.

In a worst-case scenario, such a future may come to pass through a campaign of eugenics. In the same way that early twentieth-century eugenicists believed that power over heredity constitutes humanity's greatest hope of improving itself, a genomic age eugenics movement might use the technologies of the time to attempt the same thing. If negative eugenics was once about eliminating the genetically unsound from the gene pool by sterilization, then its counterpart in the genomic age might be about preventing certain types of individuals from even entering the gene pool at all through *in vitro* embryonic genomic testing and subsequent embryo selection, also known as preimplantation genetic diagnosis, or through prenatal genomic testing and subsequent abortion. And if positive eugenics was once about the encouragement of certain types of individuals to breed to improve the stock of certain populations of humans, then in the genomic age people will look to genetic enhancement of certain traits to "improve" humans through either individualized gene therapy or permanent changes to the human germline.

Today, genetic counseling, amniocentesis, and other methods of prenatal genetic screening help parents at risk for having children with birth defects and devastating genetic diseases make informed decisions about whether to have children or to continue a pregnancy. Despite fears by critics of abortion, prenatal screening accounts for only a "minuscule fraction of all legal abortions each year in the United States and Britain." Such screening has instead "provided the vast majority of couples compelled to use it with the knowledge that their fetus is normal and with the reassurance to bring it to term." (81) But what if doctors could safely test for more than just disease status, helping parents choose selected traits for their children? What about testing for complex diseases and social behaviors with a genetic component? And what about testing for physical disabilities? Will parents take advantage of preimplantation genetic diagnosis? Will they be willing to abort their fetus based on the results of genomic analysis?

With few exceptions, this type of genomic testing offers only a prediction of what might be. A test for a complex condition or disease can only assess risk; it is not a guarantee of outcome. In this case, do we want an embryo or fetus tested? If a fetus could be tested for illnesses that have a strong environmental component, such as cancers and heart disease, then wouldn't terminating pregnancies that show a higher than normal risk for these conditions be tantamount to eugenics? What happens if scientists claim to uncover genetic influences on complex human social behaviors such as intelligence, sexuality, and aggression that we know are shaped by an endless and unpredictable array of environmental factors? Will our society sanction abortions in the case of fetuses that show a higher than average risk for these types of traits? And what about fetuses that test positive for disabilities such as blindness and deafness? Bioethicists Erik Parens and Adrienne Ashe worry that selectively aborting disabled fetuses based on genetic tests would "express negative or discriminatory attitudes not merely about a disabling trait, but about those who carry it." (82) Genomics may then serve to reinforce already existing discriminatory attitudes.

In the future, eugenics may return, in part, because the appeal of providing the best for our children may be too tempting. If parents have the opportunity to improve their children's memory, physical prowess, and resistance to disease through genetic engineering then, some might ask, why shouldn't they? Genetic enhancement technologies—the engineering of genes to make people somehow better than normal—are for now just the promises of science, but these promises may someday become reality, fulfilling the eugenic dream of creating improved humans through heredity.

Genetic imperfections are often in the eyes of the beholder.

The possibility of genetically enhancing humans faces several significant hurdles, the most challenging of which is actually making it happen. With current technology we cannot be sure of the effects of adding or removing a gene from a genome. Because most genes have more than one function, adding a gene that produces a protein that, for example, enhances memory may have unintended and potentially dangerous side effects because of the way it interacts with other genes. If enhancement would be possible, the ethical dilemmas are also tricky. The line between therapy and enhancement can, for example, be a hazy one. On the surface this distinction is meant to highlight the difference between a treatment that prevents, ameliorates, or cures a medical condition and a therapy that somehow improves upon a condition generally viewed as normal for humans. (83) Vaccinations, for example, prevent a wide variety of diseases. However, if we were to genetically modify humans to be resistant to disease, we would be enhancing the human genome. If this technology were feasible and safe, many people might choose to have their genes altered in this way. In this case the line between treatment and enhancement is blurry because "genetic vaccinations" may someday be accepted as a form of preventive medicine and not seen as an enhancement.

The line between the two becomes clearer, however, when discussing such enhancements as greater endurance or increased memory. Becuase neither is medically necessary, these enhancements might be seen as providing an unfair advantage to an individual. Because this type of treatment is not likely to be covered by health insurance and is likely to be very expensive, individuals who can afford such genetic changes would have an unfair advantage in certain sports and intellectual pursuits. Moreover, if all enhancements, including medical ones, are within the means of only the wealthy, then disparities in both health-and non-health-related genetic changes could create the dystopia described by Lee Silver that we outlined earlier in this chapter.

There may be two ways to genetically enhance humans. The first would be through genetic alterations known as somatic cell gene therapy, which would affect only the individual and not his or her progeny. In Chapter 7 we discuss the valuable medical uses of this technology. Early successes in using somatic cell gene therapy to deliver genes to treat genetic diseases suggest the feasibility of this procedure as an enhancement. Genes for potentially enhanced traits still must to be isolated, however. Currently, somatic cell gene therapy can be accomplished in both children and adults, although someday it might be possible to deliver genes in this way to a human fetus. This has already been accomplished in experimental animals. (84) Genetic enhancements could also be made to human sperm or eggs, or to embryos in the immediate postfertilization stage. Such germline changes would affect not only those treated but also their offspring. (85) If, for example, a very early stage embryo was tested and found to have Tay–Sachs disease, a genetic alteration could theoretically both cure the disease in that individual before it even expresses itself and prevent the Tay–Sachs gene from being passed to another generation.

If enhancement technologies become safe, feasible, and accepted, the eugenicists of tomorrow will most likely be people making everyday choices about themselves and their offspring. Is this a power we want? Do we have, as the bioethicist Paul Root Wolpe wonders, the "moral wisdom to know what traits are worthy of becoming incarnate in our offspring?" (86) Or do we want to place limits on enhancement technologies and their application? Because these technologies can alter the very notion of what it means to be human, we must, as a society, come to a consensus on whether genetic enhancement should become part of our lives. Medical doctor and bioethicist Edmund Pellegrino believes that the "central ethical issue underlying the use of Human Genome Project generated knowledge is making the distinction between an understanding of our bodies and an understanding of who we are as embodied beings." (87) If we ignore this distinction, we may someday find ourselves in the thicket of eugenics.

Only with an understanding of the basic ethical and policy challenges of genomics, can we be sure that the science will move ahead both safely and fairly. With the technology moving ahead so rapidly it will be difficult to sort through these challenges. The issues we have outlined in this chapter offer only a glimpse of the complicated terrain that genomics is forcing the stakeholders in this revolution, including the general public, to consider. We would all be wise to pay close attention, for these are issues that will, in one way or another, affect us all. With the assistance of programs like ELSI and the continued hard work of bioethicists and scientists, we can do our best to identify and deal with the many ethical and policy challenges of genomics. It would be a shame if our vigilance were driven not by our concern for our future, but by a fear of the science or by some genomics-related catastrophe that occurred because of our own inaction.

5

99.9%

Imagine being alive in sixteenth century Europe when Copernicus's ideas about the universe shook Europe's theological and philosophical core. It took well over a century for Nicolaus Copernicus's heliocentric, or sun-centered, theory to be accepted by the scientific and religious communities and even longer for complete public acceptance. An evolving view of the heavens altered not only the position of Earth, but also the way in which humans understood their place in the cosmos. Before the theory was established many supporters suffered for their beliefs—Galileo's ideas about a Copernican universe earned him condemnation and imprisonment at the hand of the Roman Inquisition. (1)

Or imagine reading about Darwin's theory of evolution in the 1860s and being amazed that humans, apes, and all other living and extinct species on Earth evolved from an ancient common ancestor. Resistance to Darwin's theory was immediately intense, and despite overwhelming evidence in support of Darwinism, it remains so today. (2) A poll conducted in 2001 shows that only 49% of Americans accept Darwin's account of evolution, and across the United States school districts still debate teaching evolution in biology classrooms. (3) Because ideas like Copernicus's and Darwin's altered fundamental beliefs about the nature of life and the organization of the universe, they faced ferocious resistance. Our worldviews do indeed take time to change.

If the Copernican revolution was about transforming our view of the universe and the Darwinian revolution about altering our view of ourselves in relationship to all species on Earth, then one of the most profound consequences of the genomic revolution is the way in which it is changing how humans think about themselves and one another. Despite obvious phenotypic (observable) differences, humans are remarkably alike with respect to their genomes. Our skin color, eye color, hair color and texture, sex, height, weight, and body shape may vary, but underneath these surface characteristics our genomes are all essentially the same.

Our genomes, we have discovered, are at least 99.9% identical across the great sea of diversity that is humanity. (4) This means that all humans have DNA sequences in all of our genes that are incredibly similar. Yet for centuries humans

81

Welcome to the Genome, by Rob DeSalle and Michael Yudell.
ISBN: 0-471-45331-5 Copyright © 2005 Rob DeSalle and Michael Yudell.

Despite observable differences between different peoples, our genomes are remarkably alike. All humans are at least 99.9% identical at the genetic level.

have defined themselves by their differences, sometimes manipulating the meaning of human variation to justify horrors such as war, slavery, the Holocaust, and ethnic cleansing.

Over the past 50 years, culminating with data generated by the Human Genome Project, scientists have come to understand that racial designations do not accurately reflect the biological makeup of humanity. After all, research has shown that genetic variation is greater within identified racial groups than between them, and all groups overlap when we look at single genes. (5) Without a biological foundation we are left understanding race as an historical and cultural phenomenon. Its roots lie in observable human difference, but its meanings are drawn from social circumstances. In other words, people have used race to give meanings to the social differences they believed were an extension of surface differences, and eventually they turned to science to justify these ideas. But without the support of science, what happens to the practice of dividing people by race? How will this change our thinking and our knowledge of our world? On the other hand, we know that there are grades of differences between people and populations of people that do reflect biology, and these differences may be important for understanding human evolution and to develop treatments for disease. How do we deal with the biology of human variation without giving it unwarranted social significance?

BIOLOGY TRUMPS RACE

"The concept of race has no genetic or scientific basis," NHGRI chief Francis Collins said at a June 2000 White House ceremony to mark the completion of the sequencing of the human genome. (6) At a press conference that same year Craig Venter, then president of Celera Genomics, told his audience that the data produced by sequencing the human genome have made it abundantly clear that "race is not a scientific concept." Venter noted that although the scientists at Celera could easily identify which genomes were male and which were female, they were unable to differentiate between the genomes of those who self-identified as Caucasian, Black, Asian, or Hispanic. The reason is that, as Venter emphatically stated, "On an individual basis you cannot make that determination. You can find population characteristics, but race does not exist at an individual level in science or in the genetic code." (7)

The Human Genome Project is improving our understanding of the evolutionary heritage of life on Earth and unlocking our understanding of the development and treatment of disease in humans. Nowhere in its mission was it a stated goal to undermine the biological underpinnings of race. Yet Collins, Venter, and many of their colleagues have spoken out against the notion that race is a way to scientifically describe human beings. Critics might suggest that this is simply good public relations adopted out of a fear that genomics will be tarred with the brush of eugenics. However, talk of race among genome scientists seems dictated not by personality or ideology, but by the data being generated by genomics. Furthermore, the sheer magnitude and popularity of everything genomic is sure to heighten the impact of statements about race and science on the popular imagination, on social policy, and on politics. How will the growing consensus among biologists that the concept of race is unscientific affect the fact that racial differences have a dual meaning in American society? On the one hand, racial differences can and have been used to demean and discriminate against entire groups of people. On the other hand, because of this discrimination, important social and economic policies are based on racial distinctions. Undermining the scientific basis of racial divisions has no immediate relationship to the social policies designed to repair or mitigate ongoing racial discrimination. Perhaps in the future, our knowledge of ourselves, influenced in part by a genomic view of humanity, will help eradicate discrimination and make its redress unnecessary. But for now, as a social phenomenon, race very much matters. (8) For that reason it is important to continue to sort out the social meaning of race and its impact on, for example, disparities in health care and fairness in employment.

HUMAN VARIATION

The Human Genome Project and other work in genetics reveal that all humans, on average, have gene sequences that are 99.9% identical. Just a tiny percentage, 0.1%, of humanity's genetic code accounts for each individual's unique genomic identity. But even 0.1% variation is significant, considering that the human genome consists of more than 3 billion base pairs. In every 1000-base (G, A, T, or C) stretch there is at least one difference between you and someone who is not your close relative, no matter how many physical characteristics you might share.

Sequence 1   -GCTACTACCACGGCTGCTTCGTTTGGACAAAAATAAC AGGAGGCATCCACGGG
Sequence 2   -GCTACTACCACGGCTGCTTCGTTTGGACAAAAATAACGAGGAGGCATCCACGGG

If you look closely at these two sequences of DNA, from identical regions of human chromosome 1, there is a single base pair difference. This difference, known as a single nucleotide polymorphism, or SNP, occurs approximately every 1000 bases in the human genome.

These differences, known as single nucleotide polymorphisms, or SNPs, generally have no biological consequence; they seem to be harmless variations in the DNA code. In rare cases, however, if a SNP falls in an area of a gene or a region controlling a gene's expression then it can have a functional impact on that gene by causing the production of an altered protein, too much protein, or no protein at all. By identifying SNPs scientists hope that they can target genes that may cause susceptibility to a wide variety of diseases, including cancer, diabetes, and heart disease. Scientists also believe that this type of genetic variation may also create varying responses to pharmaceuticals and environmental toxins. (9)

Genomic variation between people accounts for the diversity of physical traits that we see today among all the world's many peoples. It is these variations, visible to the naked eye, that have been used for centuries to place human populations into racial groups. Scientists now understand, however, that these physical traits are not a biologically sound way in which to divide *Homo sapiens*. We now know, for example, that sub-Saharan Africans have more "genetic variability than all other human populations combined." (10) We also know from looking around us that so-called racial groups actually contain a tremendous amount of physical diversity and that no group is defined by a set of fixed characteristics. From everyday experience we can see that no single member of a group typifies all the members of that group.

Thinking about human genetic diversity in the context of a family shows that related individuals share more of their DNA with one another. On average, parents and their biological child will have 1.5 million base pair differences in their DNA. Each parent gives the child half of his or her genome, a half practically identical with that of the parent. A child therefore has almost zero differences between his or her mother and her half of their genome and almost zero differences between his or her father and his half of their genome. Either half of a child's genome, therefore, has approximately 1.5 million differences from the opposite parent.

Differences between family members increase with greater generational distance and degrees of relatedness. A grandmother and her grandchild will have on average 2.25 million base pair differences. Biological first cousins have 2.625 million differences. Two randomly chosen, nonrelated individuals of different races, who have approximately 3 million base pair differences, are, in effect, biological second cousins, who have 2.906 million differences. Humans are effectively all, in genomic terms, biological second cousins. The erosion of genomic similarity with familial distance illustrates the nature of the human family—it is interconnected and overlapping. Making distinctions between groups of people is a difficult task, especially when the genomic relationship between all peoples is virtually the same as biological second cousins.

THE STRANGE CAREER OF SCIENTIFIC RACISM

The idea of race has a long tradition in social and scientific thought. Although the term race existed before the eighteenth century, mostly to describe domesticated animals, it was introduced into the natural sciences by the French naturalist Georges-Louis Leclerc, Comte de Buffon, in his 1749 work *Natural History*. Buffon saw clearly demarcated distinctions between the human races that, he believed, were caused by varying climates. (11) Swedish botanist and naturalist Carolus Linnaeus, the founder of modern scientific taxonomy, devised his *Systema naturae* in 1758 to classify all organisms by genus and species. Today it is still, with significant changes, the scheme scientists use to organize all living things. Linnaeus defined species as "fixed and unalterable in their basic organic plan," whereas varieties within species could be caused by such external factors as climate, temperature, and other geographic features. Human groups, or races as they were coming to be known, were also included in his classification system. Linnaeus classified four groups of humans. Unlike the other species he classified, Linnaeus divided humans by both physical and social traits. *Americanus* (Native Americans) were characterized as choleric and obstinate; *Asiaticus* (Asians) were melancholy, avaricious, and ruled by opinions; *Africanus* (Africans) were phlegmatic, indolent, and governed by caprice; and *Europeaeus* (Europeans) were sanguine, muscular, inventive, and governed by laws. (12) German scientist Johann Blumenbach's racial classifications, developed toward the end of the eighteenth century, also had a significant impact on the idea of race in modern times. Even though Blumenbach "protested rankings based on beauty or presumed mental ability," his descriptions of racial groups were imbued with a sense of White superiority. Blumenbach's racial taxonomy held that Caucasians were the racial ideal and that other groups could be understood "by relative degrees of departure from this archetypal standard." (13) The works of Buffon, Linnaeus, and Blumenbach are still, centuries later, the language through which racial difference is most often described.

Americans also contributed to the development of the idea of race. One of the United States' most distinguished racial theorists was Thomas Jefferson, who embodied a contradictory attitude toward the subject. Jefferson, who so eloquently evoked the spirit of freedom and equality in the Declaration of Independence by writing the famous words "all men are created equal; that they are endowed by their creator with certain unalienable rights," also lent his great gift for prose to scientific racism, writing in his *Notes on the State of Virginia* that the difference between the races "is fixed in nature" and that Blacks were "originally a distinct species." (14) Jefferson owned hundreds of slaves but also had as his mistress his own slave Sally Hemmings, with whom it is now widely believed that he fathered at least one child. It is an irony that Jefferson's paternity was shown in the twentieth century through DNA tests. (15)

The idea of race has had a marked impact on American history, and the United States has had its own ways of incorporating these differences into its social structure. A 1691 Virginia statute, for example, outlawed all forms of interracial marriage and punished transgressors with permanent banishment from the colony. Six of the original 13 colonies had similar laws on the books by the middle of the eighteenth century. After the American Revolution most states, slave or free, legislated against interracial marriage. (16) Distinguished American scientists articulated theories of racial difference that offered scientific justification for slavery, segregation, and other discriminatory practices. Dr. Josiah Nott, a leading physician in Mobile, Alabama, spent much of his mid-nineteenth century career touring the country, lecturing to audiences about the alleged physical anomalies of blacks and the threat he claimed miscegenation posed to whites. (17) Nott was also one of the popularizers of polygeny, a scientific discipline that espoused the idea that the human races were created in a hierarchy of separate species. (18)

For a time during the nineteenth century, polygeny lent its scientific backing to the racist cause. The experimental basis for much of this science was born in the laboratory of Dr. Samuel Morton, of whom it was written that "probably no scientific man in America enjoyed a higher reputation among scholars throughout the world." (19) Morton's work correlating cranial capacity to intelligence in so-called polygenic species was an early example of the ongoing attempt to scientifically establish an association between race and intelligence. According to Morton,

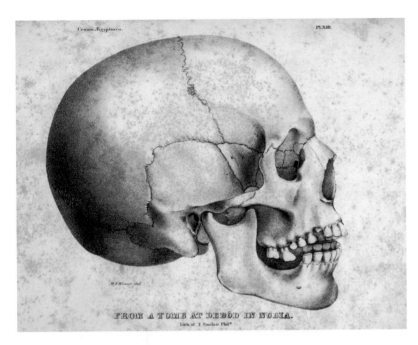

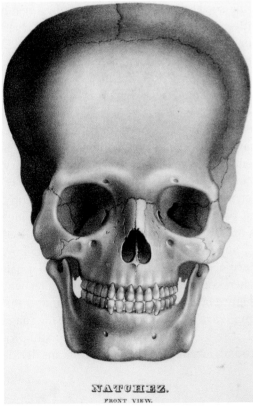

Morton's ideas about Native American and African inferiority were published in his books *Crania Aegyptiaca* and *Crania Americana*, from which both of these skulls come. Morton described Native Americans as "crafty, sensual, ungrateful, obstinate and unfeeling" and Africans as "the nearest approximation of lower animals."

Mongolians, or Asians, and modern Caucasians, or Europeans, had the highest cranial capacity, whereas Africans had the lowest.

To make this determination, Morton used tiny balls of lead to measure the volume of the skulls in his collection. More than a century after Morton conducted his experiments, the evolutionary biologist Stephen Jay Gould, with access to Morton's original skulls, recreated the experiments. Where Morton had found a hierarchy of cranial capacity, Gould could find none. Miscalculations, conscious or unconscious stuffing or underfilling of certain skulls to match his preordained conclusions, and omission of contradictory data all suggest that Morton's personal conviction about race and polygeny was "so powerful that it directed his tabulations along preestablished lines." This happened even though Morton was "widely hailed as the objectivist of his age, the man who would rescue American science from the mire of unsupported speculation." (20)

Morton's ideas were fashionable for a time during the nineteenth century, helping to develop and propagate a scientific language of difference, variations of which are still found in the literature today. But even as polygeny fell out of vogue, mostly because it was not compatible with evolutionary thought and contradicted biblical accounts of creation, other strains of scientific racism rose in its place. (21) Social Darwinism, a theory popular in the late nineteenth and early twentieth centuries argued that the social hierarchy was the natural result of a struggle for survival among the fittest. African Americans and other darker-skinned racial groups were placed at or toward the bottom of this hierarchy. (22) Specific traits associated with certain racial groups were endowed with a genetic meaning when eugenics became popular in the early part of the twentieth century. Eugenics correlated certain negative and deviant social behaviors with particular ethnic and racial populations and claimed these behaviors were hereditary and genetic. The idea that racial differences can be understood as genetic distinctions in appearance and complex social behaviors between so-called racial groups was an outgrowth of the eugenics movement. (23) Throughout the twentieth century, picking up on the work of Samuel Morton, theories of intelligence claimed IQ to be correlated with and fixed between so-called racial groups. Most recently, human sociobiology and evolutionary psychology, theories rooted in the idea that complex human social behaviors are biological in origin, have naturalized racism by arguing that a person's or group's aversion to difference is somehow wired into our genes out of evolutionary necessity: Fear of strangers was a natural adaptation in prehistoric times when competition for food and shelter was intense. (24) All of these ideas—polygeny, social Darwinism, eugenics, and human sociobiology—contribute in varying ways to the modern scientific language of race. To be sure, these are all flawed or refuted ideas. Yet their legacy still provides biological rationale, however unsound, for the ways in which humans divide themselves.

UNDERMINING THE SCIENTIFIC COMPONENT OF RACE

The way in which the United States has defined who is Black and who is White takes us to the nexus between race and science. For although it was law and custom that ultimately arbitrated these distinctions, beginning in the late eighteenth century racial divisions were hardened by the idea that differences were biological and in the blood. What is referred to as the "one drop rule" came to dominate the way in which Americans viewed race and reflect the influence that ideas about genetic heredity had over racial distinctions. The one drop rule posits that one drop of Black blood is all one needs to be Black. During slavery this rule was used to keep mixed-race children enslaved, preventing a class of mixed-race freepersons. The rule was also used during segregation to discriminate against anyone who was even suspected of having Black heritage. Even today this way of looking at race is still common and legal. As recently as the mid-1980s Susie Phipps, a Louisiana woman whose great-great-great-great-grandmother was Black, an inheritance that accounted for only a tiny fraction of her heritage, sued the state because she did not want to be labeled as Black. Federal courts upheld the one drop statute, and the U.S. Supreme Court refused to hear the case. (25)

The resiliency of the one drop rule illustrates its own contradiction, however. In the face of incontrovertible evidence showing the rule to be scientifically unsound, its power as a cultural authority remains steadfast. This has been the problem all along with race—it is a slippery category that often defies logic. Both social and natural scientists have worked to undermine the biological basis of racial categories since at least the 1930s. Yet the fact that race is considered by many to be a social and not a scientific concept often comes as a great surprise. Long before either Collins or Venter had something to say about genomes and race, many distinguished natural and social scientists maintained this position. As early as the 1930s, geneticists began moving away from the "old typological and morphological definitions of static races to [view peoples] as dynamic populations with overlapping distributions of gene frequencies." (26) Scientists were able to identify populations as differing from one another only in the relative frequency of different characteristics. As a result of this transformed view, no individual member of a race could be considered the typical member of the race, and because frequencies of alleles vary from population to population, all local populations could be considered races. Population geneticists recognized that all human populations overlap when single genes are considered and that in almost all populations all alleles are present but in different frequencies. Therefore, no single gene is sufficient for classifying humans into discrete categories. Finally, there is a tremendous amount of genetic variation in human populations, both large and small.

Human traits like skin color, hair texture, and eye shape, traits chosen throughout history to mark difference, are only part of the human façade and are a fairly arbitrary way to organize peoples. In the years following World War II many anthropologists and biologists adopted this position. Anthropologist Ashley Montagu noted that race is "one of the most dangerous myths of our time." (27) This sentiment was echoed in the "First Statement on Race" issued by the United Nations through the United Nations Educational, Scientific, and Cultural Organization, also know as UNESCO. The UNESCO Statement, published in 1950, insisted that "race is not so much a biological phenomenon as a social myth." (28) This new interpretation of race had profound consequences in Europe, which had just witnessed the horrors of Nazi eugenics, in the United States, which was in the nascent stages of the Civil Rights movement, and around the globe, where the twin legacies of conquest and colonialism affected much of the developing world. That many no longer considered race a biological fact was a revolutionary moment in Western thought, offering as it did a scientific rebuke to a longstanding belief.

THE HUMAN GENOME PROJECT AND THE FUTURE OF RACE

At the outset of the genomic revolution scientists are confident that their work has proved, at the molecular level, that race is not a biological concept. By comparing human genetic diversity in populations around the world, scientists are concluding that "the subdivision of the human population into a small number of clearly distinct, racial or continental, groups. . . is not supported by the present analysis of DNA." (29) Despite the fact that genetic studies can identify, with some accuracy, an individual's continental ancestry, this does not necessarily reflect a person's total genetic makeup. (30) This is because millennia of migration and mating have resulted in people having ancestry "from more than one major geographical region." (31) As a result, it is not possible to make simple associations between an individual's contemporary racial classification and his or her overall genetic ancestry.

Yet an important component of genomics is identifying genetic differences between individuals, and also within and between human population groups. These differences will help highlight and reveal the genetic components of disease and the biological mechanisms that cause the variable metabolism of drugs. (32) The day will come when we all enter a doctor's office with our genomes in our pockets, encoded and annotated on some type of data-storage device such as a DVD. Genomic information will be used to tailor therapies based on our own genomic idiosyncrasies. For example, Ms. Jones may show up at her doctor's office with a severe case of arthritis. According to her genome, one of six possible drugs for the condition best matches her particular genotype, ensuring better treatment of the disease and sparing her the side effects that are peculiar to her genomic profile.

Even though the speed of DNA sequencing seems to be doubling every year, DVD-Genomes will probably not be part of standard medical care for many years to come. In the meantime, however, scientists are taking the first steps toward developing more personalized medical care. The cost of whole genome sequencing is still prohibitive, making impractical the sequencing of all of our genomes, so scientists are looking to population characteristics as a first step in this process. Why not study alleles and mutations that seem to cluster in populations as a way to better understand and treat disease? Rather than using an individual profile, ancestry will be used to match a person's predicted genotype to a treatment. The challenge will be in defining these groups so that they best predict a medical treatment based on a self-reported ancestry and do not recapitulate refuted racial categories that generally do not correlate with a genotype. Such an approach needs to take into account the fact that so many Americans are of mixed ancestry. It is estimated that up to 33% of all African Americans have European-American ancestry. (33) The category of whiteness poses similar problems. Some estimates suggest that approximately 5% of those who self-identify as White have African ancestry, and White as a category includes peoples who are genetically diverse and carry different frequencies of different genetic mutations.(34) Can whiteness mean the same thing for Spaniards, Finns, and Jews? If population-based medicine is to be effective, a whole new language of human difference needs to be developed.

The advantage of studying groups or populations can be explained in the following way: If we can indeed call this or that group of people biologically similar, then it means that at some point in time there has been some barrier, be it geographic or cultural, restricting its reproduction with other groups. In such cases the gene pool will be limited by a finite group of reproductive mates, which can cause the frequency of certain rare alleles to increase. This is sometimes called a genetic bottleneck. Contemporary descendants of these populations may retain some of these genetic characteristics. Tay–Sachs disease, for example, is believed to have originated in this way. (35) But population groups do not correlate with historically constructed racial groups. None of these traits are fixed in any population.

Biologists also think that single nucleotide polymorphisms, or SNPs, can also be correlated with populations. The higher frequency of certain SNPs and/or alleles in a particular population can mean that one population will have a higher frequency of those SNPs and/or alleles than another population, making these genetic variations easier to locate. The hope is that identifiable population characteristics, such as language, geography, or culture reflect genetic characteristics and that any remedy or cure discovered for one member of the population group will most likely work for most members of that group. An example of this type of work is being conducted by deCode Genetics of Iceland. deCode, in partnership with the Icelandic government and with the participation of almost all of the Icelandic population, is attempting to capitalize on what are believed to be unique characteristics of Icelanders to show how a population group approach to genomics and medicine might work. This company believes that the people of Iceland are a good population group for studying complex genetic disorders because they have been relatively isolated for the past several centuries. The Swiss pharmaceutical company Hoffmann-LaRoche has exclusive rights to any drugs or diagnostic products that are developed from deCode's research on Icelanders. Hoffmann-LaRoche will provide free of charge to the Icelandic population any drugs or diagnostic tests produced as the result of deCode's research. (36)

The use of human population groups for scientific study is a contentious subject. The usefulness of a population group is based, in large part, on how isolated that group is. But even the Icelanders, a group thought to be reproductively isolated, may be no more inbred and genetically homogenous than most other European populations. Recent studies offer contradictory positions on this subject. (37) Despite great obstacles, humans generally do not remain reproductively isolated; there are no populations that can claim complete genetic isolation. Neither oceans nor mountains nor deserts completely stop gene flow. If this is the case, then are population groups another arbitrary measure of genomic similarity, like race? In many ways race and population are similar. They are both based on the way someone chooses to divide human beings. However, whereas racial groups originated by correlating skin color with complex social traits and prejudice, scientists try to be socially and politically neutral when delineating or naming population groups. Population groups are organized in countless ways—by geography, by migration patterns, by language, by national border, and by physical characteristics, to name just a few. Because understanding human genomic difference can help scientists unlock the genetic origins of disease, constructing representative study populations is an essential tool in the genomic age.

The Human Genome Diversity Project (HGDP) was the first large-scale genomic-era endeavor that attempted to exploit the relationship between human genomic diversity and the location of genes implicated in human disease. Beginning in the early 1990s, the HGDP sought to document human genomic diversity by collecting DNA "from widely scattered populations, which were considered to be broadly representative of the human species, preserve the DNA, and then make both DNA samples and their analyses widely available to researchers." (38) But the HGDP quickly fell on hard times. In some ways the Diversity Project suffered from bad timing and bad public relations. Even though the Human Genome Project was under way in the early 1990s, very few people were aware of its existence, and the intrusive approach of the Diversity Project—that is, to collect samples worldwide—may have moved too fast. Opposition sprang up, particularly among indigenous groups, who worried that the Diversity Project would exploit the collected DNA samples. Some of these groups expressed a fear that there was "no international protocol to protect the rights of human subjects from patenting and unjust exploitation." Others worried that the findings of the Diversity Project might even make "genetic racial injustice" a new discriminatory tool. (39)

The quest to link genes and DNA variants to disease continues today through various genomic endeavors, many of which rely on population studies to develop their data. In 2002 the NIH invested $110 million in partnership with private companies to develop a haplotype map of the human genome. The HapMap, as it is often called, will offer scientists and biomedical researchers a window into what some hope will be a treasure trove of disease-related information. (40)

Haplotypes are SNPs that are inherited in blocks of between 5000 and 200,000 bases. (41) Groundbreaking studies suggest that these blocks may be a key to understanding disease-related genomic information; some haplotypes are studied because they seem to correlate with human illnesses and they can be tested to examine their relationship to a specific disease. (42) Because SNPs can differ from one population group to another, the HapMap Project began by examining several large human populations. Those charged with choosing the population groups faced the difficult task of deciding on what basis populations should be chosen: Using traditional racial groups? Geographically or reproductively isolated groups? Or groups of the same ethnicity or national origin? In the end, the HapMap Project decided to analyze the genomes of groups that are a mix of racial, ethnic, and geographic populations and chose the Yoruba people of Nigeria, the Japanese, the Chinese, and Americans of northern and western European descent. Thousands of samples from each of these populations will be tested for SNP haplotypes, and later limited data from another 10 populations will be analyzed to see whether the genetic diversity of the Yorubans, Japanese, Chinese, and Europeans covers all of humankind. Although it is still too early to tell whether this approach to genomics will bear fruit, proponents of the project believe that it will help uncover genes associated with diseases. (43)

The choice of what populations to study and how to define them illustrates the dilemma that many research scientists face in collecting and analyzing genetic samples. Bioethicist Richard Sharp of Baylor College of Medicine and anthropologist Morris Foster of the University of Oklahoma believe that there is a contradiction inherent in such choices. They suggest that using racial or ethnic categories can help ensure "biologically diverse genomic resources," yet they also point out that such categories do not necessarily have "biological significance." In the end, these delineations are always what Sharp and Foster call "socially defined populations"– that is, their meaning is always rooted in the choices scientists make in designing their experiments. (44)

CONTINUED MISSTEPS OR A PATH TO THE FUTURE?

Many scientists continue to work diligently to show the fraudulent nature of racial categories. The work of Yale University geneticist Kenneth Kidd, for example, demonstrates that there is "a virtual continuum of genetic variation" around the world, making the delineation of races an exercise in futility. Says Kidd, "There's no such thing as race in *Homo sapiens*." Furthermore, data shows that "human genetic diversity is greatest in Africa, and the genetic heritage of modern humans is largely African." This means that "modern humans originated from a small population that emerged from Africa and migrated around the globe." (45) Research by a team of scientists from the University of Utah School of Medicine, the Harvard School of Public Health, and Louisiana State University confirms that Africans have the greatest genetic diversity of any population on the planet. This suggests that subgroups broke away from Africa to colonize the rest of the world. (46)

Yet even as many natural and social scientists work toward undermining racial categories, some areas of scientific research continue to hold fast to the past, integrating genomic data with outdated conceptions of human difference. In a worst-case scenario genomics is not a panacea for racism at all, but has an opposite effect—it ignites a genetically based bigotry similar to the American and German eugenics movements of the 1920s and 1930s. One way this might happen is that the term "population" becomes a substitute for race and the same types of prejudices once labeled racial will be transferred to populationist thinking. Moreover, the identification of disease genes or genes alleged to be found in higher frequency among certain populations may lay the groundwork for genetic discrimination a sort of end run around the idea that race is not a biological phenomenon. If, for example, a population has a higher frequency of either carrying or exhibiting a potentially dangerous allele, then are they potential targets for genetic discrimination? Will genetic discrimination become the new racism?

It would be foolish to think that science alone will change how people think about race. Indeed, despite over 50 years of work showing otherwise, the belief in the biology of race persists, with damaging consequences. Although scientists are no "more qualified than other groups of thoughtful persons to set the social and cultural ideals relating to race relations," (47) they do have much to offer to the debate, and we would all be the better for listening to what they are saying. Geneticists Kelly Owens and Mary-Claire King of the University of Washington recognize this: "Of course, prejudice does not require a rational basis, let alone an evolutionary one, but the myth of major genetic differences across races is nonetheless worth dismissing with genetic evidence." (48) Despite the genome's implicit message that we are 99.9% alike, our differences have proven to be both scientifically and medically significant. Finding a way to talk about these differences without falling back on what were misleading and destructive categories will be challenging. This new language of human difference will most likely continue to develop around the idea of populations. Although the idea of a population may sometimes overlap with more popular notions of race, the concepts do not necessarily need to be confused.

There is optimism implicit in the statements on race of Francis Collins and Craig Venter, optimism that science can and should play a constructive role in helping to define our present and future. The Human Genome Project is showing us that our genes are just a beginning for understanding human variation and an end to simplistic notions of human difference as embodied in definitions of race.

6

The Tree of Life

The cells of every living being on Earth contain DNA. From a daffodil to a California condor, from a bacterium to a giant squid, from a mushroom to a human being, the genomic alphabet A T C G is the same. The genomes of all organisms contain unique arrangements of these four bases—the information needed to sustain and cultivate life. Today's technology allows us to more rapidly read DNA sequences and identify genes in humans that are also part of the genomes of bacteria, fruit flies, worms, and mice. All species on Earth are indeed linked by their genomes, confirming what Darwin argued—that all living beings share an ancient common ancestor and thus that all organisms are related through evolution. (1) Changes in the genetic code brought about by mutations, large-scale rearrangements of chromosomes, and duplications of whole genomes are all part of the process of evolution that over time has created new species. (2) The link between genomic technologies and evolutionary theory is having a profound effect on both our understanding of the human genome and the study of evolution. Because of our knowledge of the evolutionary relationships among species, scientists can better identify genes and gene functions. And because of genomic technologies, finding these relationships can be done faster and with more precision. The story of the continuity of life on this planet is told in DNA, genes, and genomes, and this genetic history, within all of life, is being marshaled by scientists to further the field of genomics.

INHERITING A CONTROVERSY

The story of the 1925 Scopes trial is, for many Americans, an introduction to the science of evolution and the theory that humans are related to all other species on Earth. Knowledge of this trial often comes either from reading the Jerome Lawrence and Robert Edwin Lee play *Inherit The Wind*, a fictionalized account of the trial, or from seeing the film based on the play. Both the play and the film depict the backward fictional southern town of Hillsboro, dominated by simple-minded fundamentalists whose antievolution witch-hunt instigated the trial that

95

Welcome to the Genome, by Rob DeSalle and Michael Yudell.
ISBN: 0-471-45331-5 Copyright © 2005 Rob DeSalle and Michael Yudell.

is the centerpiece of the play. (3) The events leading up to the real trial could not have been further from this fiction.

In 1925, Tennessee, like other states across the South, passed a law forbidding the teaching of Darwin's theory of evolution in public school classrooms and made it a crime to teach that "man had descended from a lower order of animals." (4) Soon after passage of the law the American Civil Liberties Union took out advertisements in newspapers across Tennessee looking for a teacher willing to be a test case for the law. (5) In what remains one of the stranger publicity stunts of the twentieth century, the small eastern Tennessee town of Dayton hoped that a high-profile trial would put their town on the map. The idea to bring to Dayton a challenge to the State's antievolution laws was the brainchild of George Rappleyea, an ex–New Yorker who had come to Dayton to manage its coal and iron mines. On reading the ACLU advertisement, Rappleyea, a strong supporter of evolutionary theory, saw an opportunity to mount a challenge to the law. Rappleyea brought his idea to the

The Scopes trial was often referred to as the "monkey trial" because many believed that evolutionary theory supported the idea that humans evolved directly from monkeys and apes. Evolutionary theory instead supports the idea that humans and apes had a common ancestor at one time and that the two lineages—apes and humans—evolved separately.

town fathers, most of whom supported the new law but saw in its challenge an opportunity to bring publicity to Dayton. (6)

Dayton's civic leaders selected John Scopes, a well-liked high school science teacher and football coach, to be its teacher of evolution. Scopes, a supporter of evolutionary theory, had, as a substitute biology teacher, already assigned readings on the subject. As per his agreement with the town fathers, Scopes continued to assign his students the chapters on evolution from the classroom text. He was arrested soon after. (7)

At the time of the Scopes trial opposition to evolution took several forms. The most strident opponents believed that evolution was not consistent with the biblical account of creation. To others evolution may have been puzzling—how could humans be related to apes or bacteria? Finally, some viewed evolution with great contempt because of their commitment to fundamentalism and their political and social views. Such opposition to evolution was rooted in hostility to social Darwinism, the late nineteenth century belief that natural selection could be applied to people "with a survival-of-the-fittest mentality that justified laissez-faire capitalism, imperialism, and militarism." (8)

From across the state, newspaper editorials criticized the trial. The *Chattanooga Times* called it "the Dayton serio-comedy." The Nashville Tennessean wrote that the town fathers of Dayton were staging the trial on "the doubtful theory that it is good advertising to have people talking about you, regardless of what they are saying." (9) Many Tennesseans feared that the trial would either discredit the antievolution statute or depict the state as a hopeless backwater. Dayton's enthusiasts became so obsessed with promoting the trial that they invited the famed British author H. G. Wells to make the case for evolution even though he was not a lawyer. Despite his strong support for evolution Wells turned Dayton down. In the end, however, Dayton could not have hoped for a better marquee for the trial. William Jennings Bryan, the famed orator, three-time presidential candidate, and enduring populist leader, prosecuted the trial for the state of Tennessee. Clarence Darrow, one of the greatest trial lawyers and legal minds of his day, defended Scopes. These two larger-than-life personalities faced off in what many consider to be the trial of the century. (10)

The trial itself was primarily concerned with the legality of Scopes's arrest under Tennessee's antievolution law. That was an open-and-shut case. The jury found Scopes guilty after just nine minutes of deliberation. (11) And the ACLU's hope to use the Scopes trial as a test case to challenge all antievolution statutes never materialized—the court's judgment was deemed unappealable on a legal technicality. (12) However, the conflict over evolution and the showdown between Clarence Darrow and William Jennings Bryan gave the trial its lasting meaning. At the end of the trial Darrow put Bryan on the stand as a witness for the defense, questioning him about the literalness of the Bible and the meaning of evolution,

Clarence Darrow (left) and William Jennings Bryan (right) in the
courtroom during the trial. Temperatures in the overcrowded courtroom
soared to above 100 °. This was one of the last photographs taken of
Bryan. He died 5 days after the trial of a massive heart attack.

exposing the divide between fundamentalism and science. (13) On one side, the
prosecution asserted, "The Christian believes man he came from above, but the
evolutionist believes he must have come from below." Bryan argued that "evolution
is not truth; it is merely a hypothesis—it is millions of guesses strung together." On
the other, Scopes's defense team argued, "There is no conflict between religion and
science, or even between the Bible, accepted as a book of morals, and science." (14)

Decades later, the Scopes trial still epitomizes a philosophical divide over the
complexities of evolution. According to a 2001 Gallup Poll Americans are evenly
divided on the subject of evolution—49% of Americans surveyed believed that
"human beings have developed over millions of years from less advanced forms of
life," whereas a full 45% surveyed 'believed that "God created human beings pretty
much in their present form at one time within the last 10,000 years or so." (15)

Despite the best efforts of those opposed to evolution to cast it as something
controversial and unproven, scientific data tell an entirely different story. Evolution
rests on much more than hypotheses and circumstantial evidence. Evolution has
been subjected to rigorous scrutiny by the strict rules of the scientific method, and
now genomic data confirm what Darwin's brilliant observational skills, the fossil
record, and molecular biology have told us all along.

THE TREE OF LIFE

When cellular life began on this planet 3.5 billion years ago, the common ancestor of all life on Earth had a genome that coded for specific proteins in its cellular makeup. The genome of this ancestral organism had important functions that allowed it to survive and pass its genes to its descendants. Some of the early genes that code for basic cell function have been retained in almost all life on this planet. For example, ribosomes are present in an almost identical form in a diversity of organisms from single-celled bacterium to plants and animals. (16) These structures worked well in an ancestral cell, so why reinvent the wheel? That is the elegance and utility of evolution.

Understanding the ways in which species change and how species arise was the central innovation of Darwin's theory of evolution. On the basis of over twenty years of observation and data collection, Darwin proposed that species change over time—evolutionary time, that is. The mechanism for this change was natural selection, a process by which individuals better adapted to their environment survive and less well-adapted organisms do not. How does this process occur? Sexual reproduction, genetic mutation, and genetic recombination cause genetic variation within species—with the exception of identical twins, no two individuals are exactly genetically alike. And even identical twins can have important differences, not in their gene sequence but in how their genes are expressed. Genetic variation within a species can, under certain environmental conditions, confer advantages upon individuals with certain genes or combinations of genes. When this happens those individuals will thrive and reproduce more than other organisms. Over time, the advantages will become fixed in a population and a new species will arise. People often think of natural selection as brute competition: two members of the same species competing for the same resources with the winner of this contest passing its genes on to the next generation. An example of this would be male bighorn sheep engaging in a head-butting contest to determine who will mate with a particular female. Natural selection is rarely this dramatic. It can be as simple as favoring a particular pattern on a butterfly's wing that makes it harder for a predator to spot it than its less well camouflaged relatives. As a result, the better-hidden individual survives longer and reproduces more frequently, making a more significant contribution to the gene pool of subsequent generations.

Although natural selection favors organisms and populations of organisms that are best suited for a particular environment, it is important to realize that the environment can change, suddenly rewriting the selection criteria. An asteroid can hit the Earth or an ice age can cover the planet. In either case, the rules of the game are now different: Species or individuals within a species best suited to survive in the new environment will thrive, and those formerly best suited to the environment will lose their advantage. Because this type of change takes place over such a protracted period—often millions of years—there is something almost unbelievable about evolution. The evolutionary clock ticks at a pace almost impossible for humans to comprehend. But when you begin to look at the ways in which many species resemble one another, at the evidence found in the fossil record, and at DNA extracted from both fossilized and living organisms and see genetic similarity across a wide range of species, the biological reality of evolution becomes obvious.

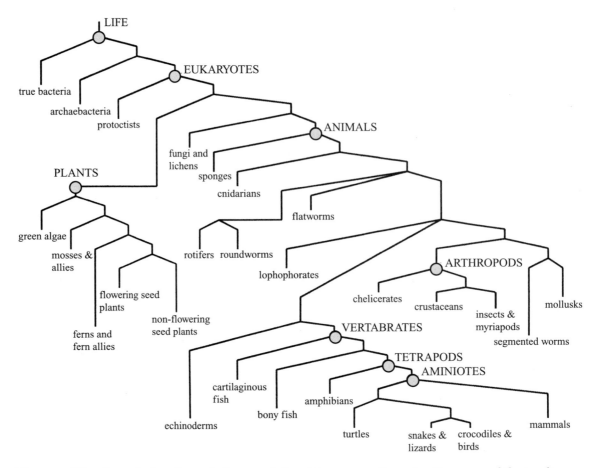

The tree of life offers a look at the evolutionary relationships among all species. You can read the tree by following the lines between each node. A node represents an evolutionary divergence, a place on the tree at which one group becomes distinct from another. In this particular tree you can see the way in which mammals are more closely related to turtles than amphibians, or how arthropods diverged from vertebrates.

The ability to compare and contrast genomes is an essential component of genomics. Thanks to new concepts and technical applications, this process is becoming faster and easier. One of the most important tools is what is known as the "tree of life"—a genealogy of life on Earth, both living and extinct. The tree, with a trunk representing ancestral characteristics, branches off into the different kingdoms of life and fills its branches with Earth's rich diversity of plants, animals, and microorganisms. Looking at the tree you can see the evolutionary relationships among all living species and their extinct ancestors. (17)

The idea of organizing life on Earth by relatedness has a rich tradition. In the eigtheenth century the Swedish botanist Linnaeus organized all living species into a nonevolutionary semihierarchical taxonomic scheme he called the *Systema naturae.* Life was organized into kingdoms, phyla, classes, orders, families, genera, and species. (18) In the *Origin of Species* (1859) Darwin integrated the *Systema naturae* with his theory of evolution in a tree of life that linked the evolutionary relationships of different species. During the last half of the twentieth century, phylogenetics, the science of building evolutionary trees, used the best techniques available to make the tree as accurate as possible. Originally, morphological data were used to organize the tree. But during the twentieth century the ways in which scientists have built the tree has evolved from looking at an organism's morphology to its biochemistry and finally to its genes. Today scientists use genomic data in conjunction with these other types of data to build the tree.

To build the tree, scientists are using the same sequencing technologies they use for the Human Genome Project. Bioinformatics, a fancy word for the use of computer science to study and compare genomes, is also emerging as one of the most important tools scientists have to understand gene sequences. Bioinformaticians write complex computer programs that can read gene sequences, locate genes, or compare sequences of the same gene in different species. (19) This is an essential component of all genomic research, but its use in building a tree of life is especially daunting because of the millions of named species involved.

The tree is a work in progress. New data sometimes force the reorganization of branches or, on rare occasions, a major rethinking of the tree's overall structure. For example, until recently the base of the tree divided into two branches, eukaryotes and prokaryotes. Prokaryotes were often inaccurately called Bacteria. Eukaryotes are all single or multicelled organisms whose cell nucleus is bound by a membrane, whereas Prokaryotes are all single-celled organisms without nuclear membranes. But in examining the ribosomal genes of Prokaryotes, Carl Woese of the University of Illinois, discovered that organisms in the group of nonnuclear membrane Prokaryotes instead fall into two very different domains: Bacteria (formerly known as Eubacteria) and Archaea. Archaea, it turns out, are more closely related to Eukaryotes than to Bacteria. Unlike Bacteria, Archaea are not inhibited by already existing antibiotics, they have a slightly different cellular structure, and their genes contain noncoding introns. Archaeans live among the most extreme environments in the world—on deep-sea vents where the temperatures hover around 100 °C, in hot springs such as Old Faithful in Yellowstone National Park, and even in the digestive tracts of cows and termites, where they produce methane gas. (20) So today we see three initial branches, or domains, of life—Eukaryotes, Bacteria, and

Archaea—diverging at the base of the tree. A more stunning result of these studies is that a previously well-accepted group of organisms—the Prokaryotes—was removed from our biological lexicon.

COMPARING SPECIES

Comparing genomic differences between species is useful as both a tool to investigate our planet's evolutionary history and as a way to identify genes and their functions. The field of comparative genomics, as it is called, identifies these genetic differences and uses this information to study our genes. An essential component of the Human Genome Project is, after all, the sequencing of nonhuman genomes including the mouse, fruit fly, yeast, and zebrafish. Through early 2004 more than 600 nonhuman genome projects were under way, and many nonhuman genomes have already been sequenced. (21)

An important goal of these projects is to compare human and nonhuman genomes to locate homologous genes. Homologous genes share a common evolutionary ancestor, have sequences that are strikingly similar, and often have the same or very similar functions. That the genes do not necessarily have identical DNA sequences is an indication of the gradual mutations that have occured during evolution, creating these slight differences.

Above are homologous fragments from β-globin protein sequences from three mammals: chimpanzee (top), human (middle), and mouse (bottom). Positions where the sequences differ are highlighted in red. Note that the chimp and human protein sequences are identical, but both differ from the mouse sequence. This is because of the very close evolutionary relationship between chimps and humans.

The identification of homologous genes reveals important and striking comparisons between human and nonhuman species. Despite outward differences, most organisms share a surprising number of genes. The microscopic bacterium *E. coli*, for example, shares approximately 9% of its genes with humans, an amazing number. When you take a look at other species, the numbers of shared genes are even more eye opening: rice—11%; thale cress—21%; roundworm—26%; baker's yeast—28%; fruit fly—45%; zebrafish—83%; mouse—89%; and chimpanzee—95%. (22) These percentages suggest two hypotheses: that there is a relationship between the genomes of all species on Earth and that species with more genomic commonality are more closely related.

HOX Genes

Different species can generally be identified by the way they look. You would not confuse a mouse and a human or a fruit fly and a zebrafish in a police lineup. Humans and chimpanzees have similar body plans but are easily distinguishable. All these species, however, share a cluster of related genes—HOX genes—that are involved in the development of the basic animal body plan. By comparing clusters of these HOX genes across the tree of life, scientists are studying how different animal body plans evolved. These homologous sequences are conserved across vast evolutionary distances, and their conservation has helped scientists locate HOX genes in the genomes of all animal species studied to date—even in the simplest of animals like sponges. The identification of the similarity of function of these genes was borne out in an elegant gene replacement study. Scientists created flies with one of these HOX genes deleted: The flies that lacked the HOX gene died.

This fruit fly has a HOX gene mutation that has resulted in it having legs where it should have antennae.

They then took the same gene from a human and placed it into the genome of the HOX gene-deficient fly: The flies with the human HOX gene replacement appeared morphologically normal and survived. (23) Mutations or nonfunctioning HOX genes can cause several physical deformities. In fruit flies a lab induced mutation of one of the HOX genes results in the replacement of the antennae with a leg.

Missing or mutated HOX genes in mice can cause either a reduction in size or the complete absence of digits and a complete absence of genitalia. Alterations in the normal structure of a HOX gene can also cause similar deformities in humans. (24)

Mice and Men

From Mickey Mouse to the mice that sometimes invade our kitchens, mice seem to have little in common with humans. But in the big scheme of life, humans are very closely related to mice and other rodents, having relatively recently (in evolutionary time) shared a common ancestor approximately 90 million years ago. That makes mice of great interest to genome researchers, who have found that almost all mouse and human genes are incredibly similar, sharing sequence order and basic biological function. For example, even though mice and humans have different-sized genomes and a different number of chromosomes, a comparison of their chromosome maps reveals remarkable sequence overlap. (25)

So at second glance mice and humans are not that different after all. Scientists have found a gene on the mouse X chromosome that can cause a crippling muscle disorder; in humans a version of that same gene, also found on the X chromosome, causes the crippling muscle disease Duchenne muscular dystrophy. (26) In another case, the human gene for a rare disorder called IPEX, also found on the X chromosome, has a homolog in mice. The homologous gene causes a disorder in mice known as scurfy, a neonatal immune disease that causes scaly skin, severe weight loss, infection, and anemia among other symptoms. Mice with the disorder die within four weeks of birth. IPEX has nearly identical symptoms in humans and is usually fatal to infants. (27)

Genomics is so important to this work because the latest DNA-sequencing methods and bioinformatics tools allow scientists to compare genes quickly between humans and other species, base pair by base pair. Because both the mouse and human genomes have been completely sequenced, a scientist studying a particular disorder can now take a gene identified in mice and quickly compare it to sequences in the human genome until a similar gene or genes is discovered. Scientists can do this by using vast genomic databases. For example, a scientist, or anyone for that matter, can go to a website called BLAST and search a database of sequences to identify similar sequences. (28) Mice are particularly important in this process because of their longstanding use as human proxies in studies of human diseases and pharmacological development and testing. From cancer to

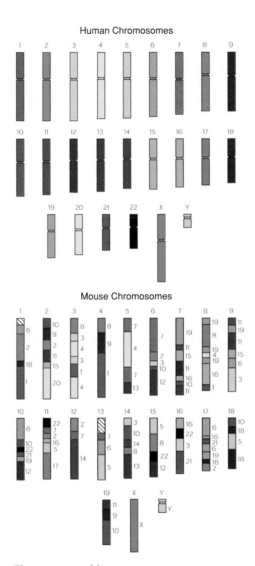

Human Chromosomes

Mouse Chromosomes

The mouse and human genomes are so much alike that the human genome can be cut into about 150 pieces and put back together in an approximation of the mouse genome. The colors and corresponding numbers on the mouse chromosomes indicate the human chromosomes containing nearly identical or homologous segments.

heart disease the mouse has been at the forefront of human medical studies. (29) The relationship between mice and humans is also based on scientists' ability to manipulate mouse genes and study outcomes. Both genetically engineered and naturally mutant mice can be studied in a controlled laboratory setting. In mice the biochemical pathways—the roads from genes to proteins to diseases—can be studied in detail in a way that complements the more limited studies that can be done with humans.

After a genetic homolog between mice and humans is located for a particular disorder, scientists can begin to study the biochemical function of that gene. For example, autosomal recessive polycystic kidney disease (ARPKD), although rare, is a cause of childhood renal failure. The disease is caused by a mutation in the gene *PKHD1*, which codes for the protein fibrocystin. A mouse homolog for this gene has been identified, so scientists can closely study the effects of this gene on mouse kidneys. The continued study of this gene and its protein product may lead to a cure for this disorder. (30)

Not So Distant Cousins

Chimpanzees are our closest living relatives. The genome sequence of the chimpanzee, it is estimated, is 98.8% identical to our own. That 1.2% difference in DNA separates us by 6 million years of evolution and several extinct intermediate ancestors. (31) In May 2002 the National Human Genome Research Institute announced that it would fund the whole genome sequencing of the chimp.

The chimp genome project, supporters hope, will offer insights into who we are in a way that other sequenced organisms cannot. (32) We know that we share most of our 30,000 or so genes with the chimp. The profound questions lie in why we are different. Can we use these differences to pinpoint the genes that give humans advanced mental and linguistic capability? And why are our close relatives not afflicted by many of the same diseases we are? Why, for example, are chimps infected with HIV but rarely come down with full-blown AIDS? Will genomic differences between chimps and humans help develop a better understanding of diseases that can lead to cures? (33)

The answers to our differences might not come only from DNA sequences, however. Some scientists believe that very few genetic differences will be found. They suggest that human and chimp biology differ in the way in which genes express themselves—that is, the way genes direct the production of proteins. Chimps and humans might have totally different ways of regulating their identical genes. The fact that our genomes are nearly identical and that there may be only relatively few gene differences also shows us that our genomes do not act gene by gene. Instead, what makes humans humans and chimps chimps is the way their genes interact with one another and the way in which environments influence those interactions. We will probably locate the differences between chimps and humans

The chimpanzee is thought to be humankind's closest relative. Its genome is 98.8% similar to the human genome.

along the interactive pathways of genes and their protein products. But our real knowledge of the differences will come from seeing the way environment and genetics together shaped how the human genome and the chimp genome evolved and function today. This is evolution at work. (34)

TAKING CARE OF THE TREE OF LIFE

Not only does genomics offer the possibility of using data from the tree of life to take care of ourselves, it also offers us the opportunity to be better stewards of the natural world. Biologists have spent a large part of the past few decades studying the DNA of many of the world's species. In the 1980s examining a single part of a single gene in the genome of an organism would have been a tedious process. By

2000 we could sequence whole genomes or at least study multiple genes of interest; we could fill many zoos and botanical gardens with species whose genes have been examined. By 2010 many of these species will have had their genomes sequenced.

We can point to many success stories about the use of DNA sequences to continue to build the tree of life and protect life on Earth. For example, scientists believed there were five species of the chambered nautilus, a deep ocean-dwelling mollusk, but DNA sequencing showed that there were actually only two species. (35) DNA evidence shows that Florida manatees, now nearly extinct, have dangerously low levels of genetic diversity, making them extremely sensitive to disease and climate change. Such information can be taken into account in designing recovery efforts. (36) Finally, researchers have used DNA to track the spread of distinct strains of the rabies virus carried by raccoons. This type of information can help animal control experts trace the origins of the rabid animals. (37)

WILDLIFE DETECTIVES

Genomic technology is being used to track and control the worldwide illegal trade in endangered species, which annually reaps hundreds of millions of dollars for black marketeers and wreaks havoc on these species and their habitats. By building DNA databases and diagnostic tools that can recognize by-products from endangered species, DNA science has become an important part of protecting wildlife.

There are more endangered parrot species than any other in the class Aves (birds). On the black market these colorful and intelligent birds can sometimes fetch more than $10,000. Found only on St. Vincent Island in the Caribbean, the St. Vincent Amazon parrot (*Amazona guildingii*) has been endangered since 1970, its low numbers linked to habitat loss, natural disasters, and its once-legal trade. To help ensure that only birds from regulated captive breeding programs are sold, conservation biologists George Amato of the Wildlife Conservation Society and Michael Russello of Yale University took samples from two of the largest captive populations of the St. Vincent Amazon parrot and identified genetic signatures, or "featherprints," unique to this group. This genomic database allows investigators to determine whether suspect parrots were bred in captivity. This information can help prosecute illegal traders and act as a deterrent to the future laundering of wild-caught birds. (38)

The Convention on International Trade of Endangered Species of Wild Flora and Fauna, or CITES, is an international protocol designed to protect plants and animals threatened with extinction. CITES works by policing the international trade in the approximately 5000 endangered animal and 25,000 endangered plant species protected under the CITES agreement. Twenty-seven species of sturgeon, a fish that inhabits rivers, lakes, and coastal waters across the Northern Hemisphere, were listed as CITES protected in 1998 because of DNA work. One of the oldest lineages

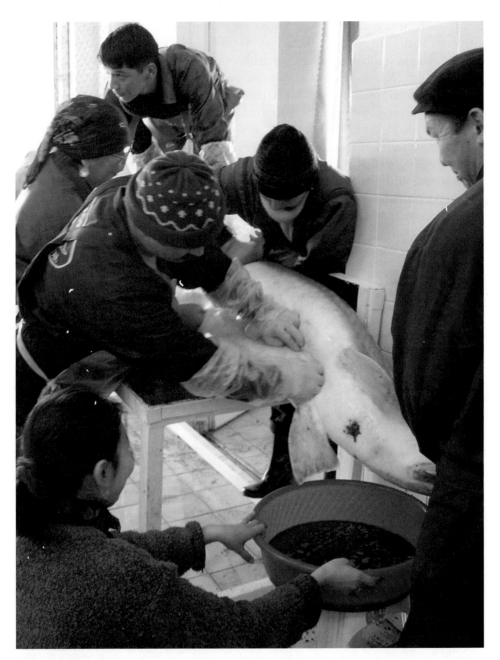

This beluga is part of an experiment to show fisheries how to extract its eggs without killing the endangered fish.

of fish, sturgeon swam with the dinosaurs 250 million years ago. Unfortunately, sturgeon are the only source of black caviar, the world's highest-quality caviar.

The consumption of these eggs, a delicacy in many regions of the world, threatens many sturgeon populations, and trade is now regulated by the CITES agreement. But it is impossible to tell with the naked eye whether the eggs come from a protected species, creating a challenge to the regulation of the caviar trade. Using DNA extracted from the caviar, biologist Vadim Burstein and his colleagues at the American Museum of Natural History were able to develop a diagnostic tool that could identify a region of DNA unique to each species using only a single sturgeon egg. Because authorities can now tell with a simple DNA test which species the caviar originates from, they can now enforce the CITES listing of sturgeons. (39)

DNA fingerprints are sometimes used to manage populations of captive animals, such as the beluga whales at the New York Aquarium for Wildlife Conservation. When two pregnant females who shared a tank with two males gave birth, researchers compared the calves' DNA to that of the males to see who the fathers were. Keepers used the results to decide which of the young should move to another tank to avoid inbreeding. (40)

DNA tests helped reveal paternity in beluga whales at the New York Aquarium for Wildlife Conservation. DNA paternity tests can be used in zoos to help avoid inbreeding.

RESCUING WILDLIFE

Until recently marine biologists recognized two species of right whales—the northern right whale and the southern right whale. The northern whales were subdivided into two populations isolated in the North Atlantic and North Pacific. Southern right whales swim across the oceans of the Southern Hemisphere. Biologists have estimated their numbers at around 8000 whales, a number that reflects the continued recovery from the devastation of nineteenth century whaling. The 300 or so individuals left in the North Atlantic have not fared as well; their numbers reflect a failure to recover from the long history of intense whaling. There are more words on this page than there are right whales in the North Atlantic. Until recently little was known about the North Pacific population, and it is believed to be in as bad shape as the North Atlantic whales, or even worse. (41)

In 2000, Wildlife Conservation Society biologist Howard Rosenbaum and his colleagues set out to better understand the genetic characteristics of right whales worldwide, hoping to use DNA data to develop conservation strategies. Because there are so few sightings of the North Pacific right whale, Rosenbaum needed to find a way to study its genes. He did it by looking back in time. Using right whale specimens from museum collections worldwide, some of the whale bones and baleen recovered from the hunts of nineteenth century whalers, Rosenbaum

DNA evidence shows that this Southern Right Whale, *Eubelena australis*, is more closely related to the North Pacific Right Whale than the North Atlantic Right Whale.

was able to isolate DNA. Speaking through their DNA, the ghosts of right whales past had something to say about their descendents, and Rosenbaum and his team listened. Because the museum specimens were catalogued with the locations of their slaughter, the data could be assembled to reflect populations in both regions.

By integrating the historical genetic data with data taken from current right whale populations, all three right whale groups could be compared. The results were not what Rosenbaum's team expected. First, instead of confirming the existence of two right whale species, Rosenbaum found three. It turned out that the North Pacific and North Atlantic populations were actually discrete species with fixed characters. The data did confirm that the group in the Southern Hemisphere was one species.

Second, in a reorganization of this tiny twig end of the tree of life, it was determined that North Pacific right whales are more closely related to southern right whales than they are to North Atlantic right whales.

Old Evolutionary Tree

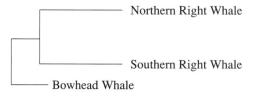

New Evolutionary Tree With Recent DNA Evidence

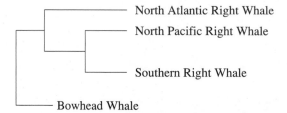

These two evolutionary trees show different ways to look at the relationship between right whale populations. The new evolutionary tree shows a third right whale species, the North Pacific Right Whale (*Eubalena japonica*), identified on the basis of DNA evidence. Bowhead whales, a closely related species, provide context for the trees.

Two populations once considered the same species were not even each other's closest relative. Finally, this information could be put to practical use. (42) After the results were published, the International Whaling Commission, the U.S. Fish and

Wildlife Service, and the NOAA Fisheries listed the new species of whale, the first newly named large whale species in over a century. *Eubalena japonica*, or the North Pacific right whale, was recognized as a new species, which meant that this whale species would be listed separately on the endangered species list. Its numbers have been estimated as low as in the tens and as high as a few hundred. It is clearly a species on the brink, but this new information may help it survive. As one of the newest members of the endangered species list, the whale will receive special protection from the United States government through species management plans and regulations. (43)

DOCTOR DNA

DNA sequences are also used in wildlife disease studies. For example, investigators in Maryland used DNA analysis to solve a murky marine mystery. A type of alga called *Pfiesteria piscicida* causes fish die-offs in the Chesapeake Bay and can also result in severe illness, including temporary brain dysfunction, in local fishermen and lab workers studying the outbreaks. Using techniques similar to DNA fingerprinting, researchers identified the DNA "signature" sequence for the species. A quick water DNA test now enables health officials to detect when *Pfiesteria* is blooming and present in large quantities and then to protect the public health by temporarily closing waterways. (44)

Putting the genome to work studying genes, enhancing evolutionary theory, and taking care of the world's species are integral parts of the genomic revolution. Synergy between evolutionary theory and genomic technology has made comparative genomics one of the centerpieces of the Human Genome Project. Discoveries made by comparing the genes of different species will someday give rise to biomedical treatments and cures for deadly diseases. Our relationship to the rest of the natural world, as shown so elegantly on the tree of life, also comes with a solemn responsibility. As stewards of our planet, humans now possess the means to use genomic data to rescue endangered species and prevent ecological disasters. Evolution and genomics do indeed work well together. And we are just getting started.

PART 3

Advancement

If the twentieth century was the century of the gene, then the twenty-first century will surely be the century of the genome, as molecular information yields incredible discoveries in both medicine and agriculture.

How far we have come.

In little more than a century, we have moved from a cursory understanding of the gene to a struggle with the intricate workings of the genome and the ways to apply this knowledge to better our lives. We do not yet know how soon these changes will make their mark, but we imagine that they will come fairly quickly. After all, just fifteen years ago technology limited our ability to sequence the human genome and today we have a completed sequence of the human genome and are sequencing the genomes of many different organisms at a very rapid pace.

In this section we look at the continued progress of genomic technologies and to a time when the promise of the genome will be realized in medical and agricultural advances. Chapter 7 looks at the breakthroughs in medical technologies that genomics is making possible and how these technologies may change the face of medicine. Chapter 8 considers the impact that genomics is having on agriculture, from genetically modified foods to the cloning of animals, and examines how these technologies are changing the farm and may affect the environment and our health.

7

The World To Come: Medicine

The achievements of twentieth-century medicine are astonishing. The discovery of penicillin in 1928, for example, made it possible by the middle of the century to routinely cure many bacterial infections that once killed indiscriminately. (1) The first success with human kidney transplantation in 1954 opened the door to all types of organ transplants. (2) And the development of chemotherapeutic and other anticancer agents allows doctors to treat cancers with increasing success. (3) At the dawn of the twentieth century Americans lived an average of 47.3 years. By century's end the average American life span had increased to 76.9 years, because of lifesaving medical technologies, a decrease in infant mortality, and improvements in public health. In 1900 the leading causes of death in the United States were pneumonia, tuberculosis, and diarrhea. (4) Better sanitation, sewage, and disease surveillance helped combat the spread of these and other infectious diseases. Public health programs promoting vaccination against a variety of diseases also saved lives. Finally, better access to health care for even the poorest patients increased the reach of medicine. Together medicine and public health did indeed make the United States a healthier nation.

Today medicine and public health are coming together in new ways. First, in the coming decades genomic advances will have a profound effect on the practice of medicine. New technologies will help move medicine away from the one-size-fits-all model of treatment, a model that is often inefficient and sometimes ineffective or dangerous. Today medical practice largely reacts to disease, with diagnosis usually the point of entry into the health care system. You don't feel well, you go to the doctor, you are diagnosed, and you are, if a therapy exists, treated, and, you hope, cured. In the age of genomics medical practice will be different. Greatly improved knowledge of human health and disease, the creation of medical care that is personalized to an individual's genome, and the responsibility to interpret and

117

manage billions upon billions of base pairs of genomic information will all be part of the coming transformation in medicine.

Genomic technologies will also become an important part of public health services. A better understanding of the origins of human diseases, for example, will allow people to take more informed preventive measures. Genetic testing will allow for the identification and possible prevention or treatment of genetic diseases. Finally, public health efforts will help elucidate the gene-environment interactions that are responsible for almost all human diseases. (5) For example, the new fields of environmental genomics and toxicogenomics will enable scientists both to identify the genes that make humans vulnerable to certain environmental exposures and to better characterize the toxic substances that pose a threat to human health. This information can ultimately improve our understanding of the relationship between environmental exposures and human disease. The changes that genomics brings to both medical and public health services should help craft a health care system that improves human health by preventing, predicting, and treating disease with increasing accuracy and effectiveness. (6) The impact of these technologies on medicine and public health worldwide is so far a very different story. The introduction of genomic technologies into developing countries will be hindered by several factors including poor medical infrastructure and a lack of commercial interest in treating many of the diseases that strike in the Third World. (7) In 2002 the World Health Organization endorsed the creation of a Global Health Research Fund that will facilitate the integration of genomic tools into developing countries. World Health Organization Director General Dr. Gro Harlem Brundtland believes that genome technology "has the potential to allow developing countries to leapfrog decades of medical development and bring their citizens greatly improved medical care." (8) Delivery of lifesaving genome-based medical technologies to the developing world should be a yardstick with which to measure the ongoing success of genomics.

Despite the great promises of genomic technologies we must remember that we are just getting started in this endeavor. Nobel laureate Harold Varmus, a pioneer in the study of cancer genetics and now President of Memorial Sloan-Kettering Cancer Center in New York City, reminds us that the sequencing "of the human genome did not mean that the practice of medicine would be abruptly and radically transformed." (9) Genomic medical technologies will, to be sure, affect us all in one way or another in our lifetimes, but the real fruits of this research will develop over time. We worry that if the promises of the genome are not met immediately, people will be disappointed. "Wasn't all that information from the genome supposed to cure my disease?" people might ask. In the coming years as more and more people benefit from genomic technologies, that attitude is sure to change. In the meantime, we can ready ourselves for the arrival of genomic medicine by learning more about these new technologies and the ways in which they will affect our lives.

ENTERING THE AGE OF GENOMIC MEDICINE

Thus far we have discussed several human genetic disorders that are under the control of a single gene. For example, sickle-cell anemia, despite its devastating effects, is caused by a single genetic mutation. Other deadly diseases like Huntington disease and Canavan disease are also single-gene genetic disorders whose origins are now known. The vast majority of all human disease is, however, polygenic in origin—that is, more than one gene is involved in the development of the disease. Furthermore, with polygenic diseases the environment generally plays a fundamental role in disease development. Kenneth Olden and Samuel Wilson of the National Institute of Environmental Health Sciences liken the relationship between genes and the environment to a loaded gun and its trigger. Only when a trigger is pulled can a loaded gun do harm. Likewise, an environmental exposure is needed to trigger one or a combination of susceptibility genes that can cause disease. (10)

Understanding the genetic origins of both simple and complex diseases is no easy task. The hunt to locate the Huntington gene, for example, took over 10 years of collaborative research. Located on chromosome 4, the gene causes Huntington disease when a small segment of the gene is spelled over and over, more than 38 times, almost like a stutter. Normally, the bases C, A, and G can repeat up to 32 times in this gene. When the repeat occurs too many times, Huntington disease results. More repeats correlate with earlier age of onset of the disease. This "molecular stutter" is devastating, and Huntington disease patients suffer terribly; the disease is untreatable, and neurological symptoms often appear in a patient's thirties. (11) Onset of the disease causes the degeneration of the central nervous system, and death follows within 10 to 15 years.

Medical conditions like most cancers, heart disease, obesity, and allergies, on the other hand, are the result of the interaction of many genes. Determining the genetic basis of a complex disorder is thus a daunting task, but even if a disease has a complex origin, knowledge of its genetic component(s) may help in finding a cure. Even though more than one gene may be involved in a disease, it does not necessarily mean that a cure must repair all the genetic errors involved in the disease. This is the reason that genomic information is so important to the development of new drugs. Knowledge of a gene can lead to a therapeutic intervention that could repair, replace, inhibit, or supplement the activities of a gene, an RNA molecule, or a protein.

One of the most important tools to emerge from the genome project is the microarray, a device sometimes also known as a gene chip. In any cell, under different circumstances, only a fraction of the 30,000 or so human genes are active. The microarray allows researchers to look at the activity of many individual genes and also to look at the interactions of as many as thousands of genes at a time. This tiny genomic instrument, which can be as small as a quarter, may appear modest, but it is revolutionizing genomic research and medicine. Microarrays have many applications, from locating genes that are potential targets for pharmaceutical development to pinpointing the genetic roots of diseases to studying genetic responses to environmental toxins.

This tiny gene chip is helping scientists sift through the enormous amount of data generated by genomics.

Microarrays will improve the quality and scope of genetic tests by allowing doctors to assess disease risk for multigenic traits. Pharmaceutical companies may use this technology to streamline drug development so that patients one day receive treatments tailored to their own genomes. Advances in robotics, chemistry, and engineering led to the development of the microarray, but the Human Genome Project made it a feasible and useful technology. Microarrays need a great many genes to be of practical use, and their utility depends on the genome project and its rapid identification of genes in human and other genomes.

To understand how a microarray works, we need to remember how genes work. (12) In Chapter 2 we discussed the central dogma of molecular biology: DNA is transcribed into messenger RNA (mRNA) that is in turn translated into a protein. This sequence of cellular events is key to understanding a microarray. It is the presence of mRNAs, the intermediary product between genes and proteins, that shows which genes are active in a cell at that moment. A microarray allows scientists to observe this process and record which genes are on and off in a cell.

The typical microarray works like this: Scientists first choose a specific type of cell they want to study—for example, colon cancer cells, cells from heart tissue, or blood cells. A computer selects short stretches of DNA sequences from all known genes in the human genome and places them onto the microarray, which is generally either a simple glass microscope slide or pieces of nylon affixed to plastic. A close look at the array reveals tiny dots on its surface. Each dot represents a segment of a gene. A single gene is, for accuracy, typically represented by several dots, and the entire matrix of dots typically represents thousands of different genes. At each tiny dot there are copies of identical strands from a single gene. With the help of a robot, up to tens of thousands of these DNA stretches can be precisely affixed to a single microarray. Separately, mRNA, which is isolated from the cell type under study, is prepared and labeled with a fluorescent dye. Because mRNA from a specific cell type is used, we know that it is being produced by genes that are currently active in the cell type we are studying. The mRNA collected represents the transcribed material from a particular type of cell.

Once the mRNA has been isolated and prepared[†] and has had a glowing fluorescent tag attached, the solution of prepared mRNA is washed over the gene-spotted array. The temperature of the microarray is raised to stimulate hybridization of the DNA strand on the microarray and the prepared mRNA strands that were washed over the array. When a strand of the prepared mRNA in the wash and the DNA in a spot are complementary, the fluorescently labeled prepared mRNA will hybridize and "stick" to the microarray. In other words, the prepared mRNA finds the genes on the array from which they were originally transcribed, forming a double helix. A special machine then magnifies the chip, allowing the thousands of gene spots on the array to be read for the presence of fluorescence. If a spot on the array does not glow, then the gene is not active, or is "off." However, if a spot glows, then the gene is active, or "on," in the cell under study. The brighter the spot glows, the more active the gene is. We know this because we can correlate gene activity with the amount of mRNA found in a cell.

Microarrays can be used to compare some types of cells in different states. For example, microarrays have proven very useful in comparing cancerous and normal cells. Cancer cell microarray studies begin in much the same way as the procedure just described, with one important difference. Because we are comparing two different states of the same cell types, separate batches of mRNA are prepared from the cancer cells (labeled with red fluorescence) and from the normal cells (labeled with green fluorescence). Next, the two pools of prepared mRNA are simultaneously washed over the array. The genes dotted on the surface of the microarray will glow green if they are active in the normal cells and red if they are active in the cancer cells. If the genes are active in both cancerous and normal cells,

---

[†] (prepared mRNA is actually DNA copied from mRNA)

Resembling a stained glass window, this representation of a glowing microarray display demonstrates the state of the genes in a normal cell and a cancerous cell. The green spots are genes active in a normal cell, the red spots in a cancerous cell, and the yellow spots in both cells. The brighter the glow, the more active the gene.

the spots will glow yellow because of the combination of red and green fluorescence at that spot. If a gene is not active in either cancer cells or normal cells, there will be no fluorescence on that spot. Studies are now underway to determine the accuracy of microarrays in predicting whether a cancerous tumor will metastasize. Studies on breast cancer patients have shown that a microarray can help predict whether the cancer will spread and whether it will respond well to standard treatment. The genes that designate poor prognosis on a microarray may eventually prove useful. By zeroing in on these genes, researchers may have better targets for which to develop drugs. (13)

Diffuse large B cell lymphoma (DLBCL) accounts for approximately 40% of all cases of non-Hodgkin lymphoma, a cancer of the lymphatic system. Sufferers of this disease can have very different clinical outcomes. With the standard treatment fewer than 50% of patients are cured. But microarrays have helped identify three distinct forms of DLBCL, and scientists now believe that these different forms are actually different diseases and should be treated as such. The microarray enables researchers to classify patients by DLBCL subgroup and predict the likely response to the standard chemotherapy, to which only one of the three subgroups responds with a high success rate. This technological improvement allows doctors to save nonresponsive patients from a debilitating round of chemotherapy and to seek alternative and hopefully more successful treatments. (14)

The promise of genomics and medicine also lies in the ability to locate genes involved in human disease and in turn to study these genes to help find cures. Microarrays are an important part of the genomic arsenal that enables scientists to sift through the approximately 30,000 genes in our genome in the ongoing search for the biological origins of disease. The path from gene to drug development is illustrated by the discovery of a drug to treat chronic myelogenous leukemia or CML, a journey of almost 40 years. In 1961 scientists discovered a chromosomal abnormality and consequential genetic mutation—the fusing of the Bcr and Abl genes—associated with the cancer CML. (15) Forty years later, in 2001, the FDA approved a drug known as Gleevec for the treatment of CML. (16) Gleevec inhibits the production of a by-product of the gene that causes the uncontrolled production of white blood cells in CML patients. Unlike chemotherapeutic agents that also target healthy cells, Gleevec is what is known as a rational drug; that is, it interacts directly with the cancer and not with all cells in the body. Gleevec has shown unparalleled success in treating this once deadly form of leukemia. (17) Although the disease is potentially curable with a stem cell transplant, fewer than 30% of patients have suitably matched donors and only 5–20% have a positive response when treated with interferon, an alternative treatment for the disease. Patients with an advanced form of CML who have been treated with Gleevec are showing a remarkable response. In one study, after 18 months of treatment, 88% of patients showed no signs of disease progression and 94% of patients were alive. Furthermore, only 2% of the Gleevec patients discontinued treatment because of side effects, and no patients died as a result of taking the drug. (18) It is hoped that in future it will no longer take 40 years to exploit the relationship between genes and therapies. Microarray and other genomic technology can quicken the pace of drug discovery by helping scientists locate genes more rapidly, by improving our understanding of the genetic etiology of disease, and by locating targets for drugs.

Scientists are working hard to exploit the important connection between gene and therapy. With genomic knowledge in hand, researchers hope to develop drugs that have less toxic side effects and that precisely target the biological or genetic cause of the disease. In 2001, for example, researchers at the National Human Genome Research Insitute (NHGRI) discovered a new genetic component in some types of Parkinson disease. The protein α-synuclein is abnormal in Parkinson disease patients. Researchers at NHGRI hope that by understanding the genes responsible for this abnormal protein, they may someday be able to treat Parkinson disease by either fixing or controlling the expression of this protein. (19)

The Human Genome Project has set in motion other initiatives that promise to complement and enhance molecular biology and its application to medicine.

For example, although there may only be 30,000 or so genes in the human genome, those genes could direct the production of approximately 500,000 proteins. (20) Scientists believe that drug development will be greatly improved by understanding the human proteome, the entire collection of human proteins. A sharp rise in proteomics research since the sequence of the human genome was completed marks an important new direction in molecular research. The study of the structure of proteins, their interaction with one another, and their presence in different cells may help lead scientists to new targets for drug therapies. (21) A coordinated Human Proteome Project, much like the Human Genome Project, is in the planning stages. (22) Yet despite all these efforts, advances in medicine will most likely come slowly. Clinical trials alone take, on average, between five and seven years before a drug is approved for the public. That time does not include preclinical laboratory work that we know can take even longer.

FORTUNE TELLING

If someone could tell you your medical future, would you want to know it? Might you want to limit the information that tests could reveal? Would you want to know what illnesses you might someday develop? Genetic testing offers the possibility, under certain circumstances, of answering some of these life-and-death questions.

Although genomic technologies will someday quicken the pace of drug discovery and open the door to new types of cures, the impact of genetic testing is already being felt. Genomic data are helping unveil associations between genes and the diseases they either cause or contribute to. Once a gene for a particular disorder has been identified, it is then possible to develop a genetic test for that disorder. As of October 2003, there were 993 genetic tests that identify genetic causes of disease, of which 645 are available for clinical use. The rest are currently used for nonclinical research only. (23) Microarrays and DNA sequencing are some of the technologies being used in genetic testing.

There are several circumstances under which a genetic test is offered. Today the most commonly used tests are those that diagnose an illness and those that help people decide whether to have children or how to manage a pregnancy. Diagnostic tests are generally used to confirm symptomatic disease. For example, DNA analysis might be used to see whether a child with chronic pulmonary infections has cystic fibrosis or a young boy with muscle weakness and motor control problems has X-linked muscular dystrophy. (24)

Diagnosis of a genetic error has great value when it can be used to identify treatable diseases. Phenylketonuria, or PKU, is a genetic disorder that causes a deadly metabolic malfunction. The approximately 1 in 12,000 babies born each year with the disease cannot metabolize the amino acid phenylalanine, a common component of many foods. If the disease is untreated, phenylalanine by-products build up in blood and tissue and cause severe mental retardation. A screening test was developed in the early 1960s to help detect the disorder. If the test is positive for PKU, doctors prescribe a phenylalanine-restricted diet and children with the disorder can grow up normally. Today, in the United States and many other countries, all newborns are screened for the disease. (25)

Some diagnostic genetic tests can reveal presymptomatic disease years before its clinical manifestations. For example, DNA testing can identify individuals who will develop Huntington disease many years, if not decades, before symptoms appear. However, because there is no known cure for Huntington disease, the information obtained from the genetic test can be life-changing and frightening. Those identified as having the illness may not want to risk having children because there is a 50% chance in each pregnancy that the child will inherit this incurable disorder. People may also plan their lives differently. Someone who knows that he may become gravely ill in his thirties or forties may make different choices in life.

Tests can be used to determine whether an individual is a carrier for a particular illness. If both members of a couple are found to be carriers, genetic counselors can help the family understand their risks for having a child with that problem. Screening programs for at-risk parents help assess whether a genetic test is necessary. For example, parents who are carriers of the gene for Canavan disease, a devastating neurological disorder, risk passing the disease onto their children. Canavan is a recessive disease, and, as in most recessive diseases, the carrier parents exhibit no symptoms. A child must inherit the defective gene from both parents to develop Canavan disease. If only one parent is a carrier, then there is a 50% chance that the child will also be a carrier but not develop the disease. If both parents are carriers of the Canavan gene, then there is a 25% chance that their child will be born with the disorder, a 50% chance that the child will be a normal carrier, and a 25% chance that the child will not carry the gene at all. Parents who are both carriers of Canavan, armed with such knowledge, may or may not choose to have children. (26) Other recessive diseases include cystic fibrosis, Gaucher disease, and sickle-cell anemia.

Genetic counselors are on the front line of advances in genetic testing. Their job is to work closely with people being tested to ensure that before testing they are informed of all the ramifications of a genetic test, that they are given the appropriate genetic test once they make an informed decision to proceed, and that the testing information is correctly interpreted. The mutations for diseases like Tay–Sachs and sickle-cell anemia run in families, thus making the results of a genetic test of interest to other family members. (27) In such cases a genetic counselor may urge a patient to share the results of his or her genetic test with those individuals. As genetic tests become increasingly common in medical practice, the currently limited number of genetic counselors will need to increase and primary care physicians will require better training in interpreting and explaining genetic tests. (28)

Testing for Genetic Diseases

Disease	Type of test	Incidence
Breast and ovarian cancer (one hereditary form)—Particular genetic mutations leave individuals with increased risk of develeoping breast, ovarian, and other cancers.	Predictive—Molecular genetic test for presence of mutation in BRCA 1 and/or 2 genes.	Rare—Mutation accounts for less than 5% of breast and ovarian cancer cases. The risk for developing breast cancer for women who have either mutation is 60–80% over a lifetime. A number of viable treatment options exist, but success is often dependent on stage of disease at discovery.
Cystic fibrosis—A serious disease that affects multiple organs and systems. Characterized by thick mucus buildup in the lungs, leading to lung infections and problems with the pancreas that in turn cause digestive and nutritional problems. Certan CF mutations cause male infertility. Life expectancy is shortened.	Carrier, prental, and diagnostic—Molecular genetic test for mutations in CFTR gene.	Rare—Affects 1 in 2000 newborns. Improved treatments have increased the life span of CF patients, but death ususally results by about 30 years. 1 in 31 Americans are carriers of the disease.
Down syndrome—Also known as trisomy 21. Caused by an extra chromosome 21. Usually results in some degree of mental retardation and physical deformitites.	Prenatal and newborn screening—Examination of chromosomes for extra chromosome 21.	Most common human birth defect. 1 in 660 newborns are affected. No known cure. Advances in treatmeent of some of the disorder's symptoms have resulted in greatly increased life span for Down patients, who can live more productive lives with educational interventions.
Familial adenomatous polyposis (FAP)—A malignant disease of the colon. Hundreds to thousands of colon polyps develop, invitably leading to colon cancer.	Diagnostic—Molecular genetic test for mutation in APC gene.	Rare—Cure with colectomy, removal of colon.
Fragile X—Affected individuals suffer from moderate to severe mental retardation and some facial and behavioral abnormalities.	Prental and diagnostic—Molecular genetic test for mutations in FMR1 gene.	Rare—Affects 1 in 2000 newborns. No known treatment.
Huntington disease—A degenerative neurological disorder that affects brain function and inevitably leads to death.	Diagnostic—Molecular genetic test for mutation in the HD gene on chromosome 4.	Rare—If you have a parent with the disorder you have a 50% change of having Hungington disease. No known cure.

Continued

Disease	Type of test	Incidence
Phenylketonuria (PKU)—Intolerance to amino acid phenylalanine. Results in severe mental retardation if untreated.	Carrier and newborn screening—Newborns are given a blood test and molecular genetic testing of the PAH gene. At-risk couples can be tested for carrier status.	Rare—Affects 1 in 10,000 newborns. Disease is treatable with identification of disorder and dietary restrictions.
Sickle-cell disease—A group of diseases that cause a change in the shape of red blood cells so they look like sickles, causing severe anemia. Can lead to severe organ damage and sometimes death.	Diagnostic, carrier—Molecular genetic test for mutations in HBB gene found on chromosome 11.	Rare—Sickle-cell diseases include sick-cell anemia and beta thalassemia, among others. The genetic mutation that causes sickle cell anemia was identified in 1957. There is still no known cure for the disease.
Tay–Sachs—Degenerativae neurological disease characterized by loss of motor skills in infant between 3 and 6 months. Complete neurodegeneration leads to death before 4 years.	Carrier, prental, and diagnostic—Molecular genetic test for mutations in HEX-A genes on chromosome 15.	Rare—1 in 250 people are carriers of the disease. Incidence is known to be significantly higher among Ashkenazi Jews, French Canadians, and Cajuns. No known treatment.

This table highlights several genetic diseases with both complex and simple etiologies, and the type of tests offered for them.

Genetic testing is also regularly used to determine the genetic status of a fetus. Intrusive and sometimes risky procedures like amniocentesis and chorionic villus sampling can genetically test a fetus. Such tests, now common among women pregnant with their first child after the age of 35, look for chromosomal abnormalities such as Down syndrome. (29) Advances in genomics will, however, offer the possibility someday to test fetuses for more complex disorders. Bioethicists, theologians, and policy makers are debating the morality and practicality of such tests (see Chapter 4).

Predictive genetic testing is, in its most basic sense, fortune telling. A predictive genetic test looks at one or many of your genes and assesses your risk of developing a particular illness. There are two classes of predictive tests—those for disorders caused by a single gene and those for complex genetic disorders that involve multiple genetic and environmental factors. With disorders of a single gene, the test is either negative or positive. A person either has the disorder or doesn't; for example, it is certain that an individual who has more than 38 CAG repeats in their Huntington gene will some day develop Huntington disease. Complex diseases are different. Depending on the nature of the disease, the risk of developing the disorder can range anywhere from zero—the individual does not have a genetic risk—all the way to a 100% certainty of developing the disorder.

Because predictive testing for complex genetic disorders is about probability, it is therefore only an estimate of risk. In predictive testing for such disorders an individual can have a very high probability of developing a disorder and never get sick; some individuals will beat the odds. On the other hand, test results can show an extremely low probability of getting the disease and you can still get it. The results of predictive testing can be difficult to decipher because most diseases are caused by many genes interacting with environmental triggers. Moreover, a disease is not necessarily genetic because it can be genetically tested for. Many conditions have multiple causes, both genetic and nongenetic. For example, the genetic test for a mutation in the genes BRCA1 and BRCA2 indicates only a probability of developing breast cancer and not a definitive outcome. Women with either of these mutations have between a 60% and an 80% chance of developing the disease. The mutation in these genes accounts, however, for fewer than 5% of all breast cancer cases. (30) So even if you test negative for BRCA1 and BRCA2, you can still get breast cancer.

In cases like the BRCA1 and BRCA2 tests, where the results offer insight only into the probability of developing the disease, would you want to have the test? Women who test positive may benefit from closer medical surveillance and possible lifesaving early detection, or they may choose to undergo prophylactic mastectomies. On the other hand, they may experience stress and anxiety from knowing that they face an elevated risk of breast cancer. What if a genetic test revealed that the risk was only slightly elevated? Or what if knowing the risk could lead to lifestyle changes that decreased or eliminated the risk? What if the lifestyle changes were too compromising? Would you still want this information? Finally, what if this genetic information revealed not only your future but the genetic future of other members of your immediate family? Would you share this information with them?

Genetic testing not only offers potentially lifesaving and diagnostic medical tools, it also presents, like most genetic information, a host of medical challenges and moral dilemmas. We know that none of our genetic tests will be perfect. It's wrong to even think of genetic tests as things to "pass." Most of us would want to avoid finding out that we have a variant of a gene strongly associated with a serious disease. But in a sense we would all fail genetic tests in that none of us have the most "healthy" form of every single human gene. We all possess gene variants less well adapted than others. In fact, it is not really possible to define the healthiest form of a gene, because the value of a gene only has meaning in relation to a shifting and changeable environment. So what do the results of a gene test mean, and how will we cope with them?

Ashkenazi Jews, once victims of the Nazi campaign for genetic purity, face a greater genetic risk of Tay–Sachs disease as well as some forms of colon, breast, and ovarian cancers than the general population. In 1998 Jewish groups met with Francis Collins, the director of the National Human Genome Research Institute, and Richard Klausner, the director of the National Cancer Institute, to express their "concern over the lack of legal protections against genetic discrimination." Collins believes that genomics will show that Jews are no more predisposed to genetic diseases than other population groups. The "silver lining," Collins said, might be that Ashkenazi Jews "may be the first to benefit from drugs and other therapies that will derive from the work." (31)

The ultimate genetic test would be a whole genome scan capable of assessing genetic risk for a range of illnesses. Such a test may become a reality as the cost of sequencing a whole genome becomes less prohibitive. As head of Celera Genomics, Craig Venter led the effort to sequence the first draft of the human genome in 2000. Today, as head of the Center for the Advancement of Genomics, Venter hopes someday to develop technologies that reduce the price of sequencing a genome to as low as $1000. By providing people with their genetic information Venter hopes that this technological leap will some day "apply the genome to people, not to just put it on a pedestal." (32) Venter's move should have a significant impact on speeding up the sequencing of personal genomes. These benefits are sure to have an important impact on medicine by identifying an individual's genetic susceptibilities, but for now we still know so little about the genetic origins of most diseases. Until that changes, people might come to fear so-called genetic irregularities that can be identified by a $1000 genome sequence but turn out to have no significant biological impact. Policy makers are aware of the potential gap between test and therapy. Francis Collins and his colleagues at the NHGRI argue that "making certain that genetic tests offered to the public have established clinical validity and usefulness must be a priority for future research and policy making." (33)

Hundreds of genetic tests are now clinically available, but most identify diseases that are generally rare and untreatable. (34) The tests will remain most useful as diagnostic tools or as a way to determine carrier status. Until treatments for these diseases are developed, these tests can have only limited impact on the practice of medicine. Predictive genetic tests can, however, serve an immediately useful purpose, especially for diseases whose risks can be reduced through preventive measures. If a test shows a higher risk for heart disease, then at-risk individuals may be able to prevent the onset of the disease through diet changes and exercise. This aspect of the genomic revolution will be an effective medical tool if doctors and genetic counselors are able to communicate successfully to individuals what their risks are, how those risks contribute to the development of disease, and how risks can be avoided. But the danger of predictive testing may outweigh the benefits. A positive predictive genetic test could lead to medical interventions that one genetic testing expert calls "costly, unnecessary, ineffective, or even harmful." (35)

Genetic tests can also be used to determine how genetic variation affects individual reaction to drugs and to environmental exposures. The following two sections explore how these types of genetic tests will affect our lives.

PHARMACOGENOMICS

From over-the-counter remedies to doctor-prescribed medications, reactions to a drug can greatly vary from person to person. Drugs are designed for the average person, so sometimes they work effectively, sometimes they are ineffective, and other times they can have dangerous and even life-threatening side effects. A study conducted in the mid-1990s showed that adverse drug reactions "ranked between the fourth and sixth leading cause of death in the United States." (36) The new field of pharmacogenomics hopes to eliminate these adverse reactions and treatment failures by targeting drug treatments to the individual. How will scientists do this?

It would be so easy to treat diseases if all humans had the same genome. The same drug could be administered to everyone to cure a given disease. But human genetic variation is an extraordinarily important part of both what makes us human and what allows us to survive. Without variation within our genomes we would all have the same disease susceptibilities, and thus a single disease could extinguish humanity. Fortunately, human genomes are, on average, only 99.9% identical. This seemingly slight variation protects our species, but it also makes the development of drugs to treat diseases very difficult. Genes can have slightly different sequences and still function normally, so differences generally remain hidden and harmless. Scientists are discovering, however, that these differences can have a significant effect on the way in which individuals metabolize drugs and the way in which a disease develops. By studying these differences scientists are improving patient care by identifying patients who cannot metabolize certain drugs because of a genetic variation that causes changes in their drug metabolizing enzymes and by developing drugs that avoid metabolic pathways that are identified as causing an adverse drug reaction or treatment failure. (37)

The anticlotting drug Warfarin, one of the most widely used drugs in the world, offers a good example of the way in which pharmacogenomics can work. Warfarin is metabolized primarily by the enzyme CYP2C9. Patients who are deficient in CYP2C9 can suffer dangerous and deadly bleeding episodes because of this deficiency and should take an alternative drug or a lower dose of Warfarin. A simple genetic test can help identify people deficient in CYP2C9 and help avoid dangerous side effects. (38)

Cancer therapy is another area in which pharmacogenomics holds great promise. Because cancer cells are not very different from normal cells in your body, most anticancer agents cause unwelcome secondary effects in noncancer cells. Anticancer drugs can be extremely dangerous and unpredictable because there is often not much difference between a toxic and a therapeutic dose. (39) Drug responses have been studied in the cancer acute lymphoblastic leukemia and have shown that patients with a particular genotype have a better chance of being cured than other patients. (40) Knowing these distinctions allows doctors to tailor a therapy to a particular genotype in an attempt to avoid side effects and to enhance treatment.

The whole genome sequences generated by the Human Genome Project and the Celera genome effort offer a glimpse into the genomes of only a few individuals. Although these sequences are shedding light on the nature of the human genetic code, other genomics initiatives are building databases of genomic variation. Pharmacogenomics researchers are interested in this variation because it can help identify one of the ways in which genetic variation influences drug response. Single nucleotide polymorphisms (SNPs) are an important and simple aspect of genomic variation. They are single base pair differences that can be found when looking at the same spot in different people's genomes. SNPs are fairly common, generally occurring once out of every thousand base pairs in the human genome. Identifying SNPs that may play a role in the treatment of human disease is an important component of pharmacogenomic research. The SNP Consortium, an ongoing joint venture of several pharmaceutical, biotechnology, governmental, and philanthropic institutions, has created a map of the approximately 1.8 million SNPs they have thus far been identified in the human genome. (41) The Consortium releases its information to the public without any intellectual property restrictions in the hope that it will promote pharmaceutical research. (42)

There are several ways in which pharmacogenomics can be effective. The first is studying populations for pharmacogenomic variability: Can we identify populations sharing genetic characteristics that can cause some type of pharmacological failure? If we can, then an individual's ancestry could help determine the choice of drugs for his or her treatment. In Chapter 5 we spoke about a contradiction in the Human Genome Project. On the one hand, genomics has shown that race is not a biological phenomenon and that any two randomly chosen, unrelated individuals of different races can, on average, share more genetic material than two randomly chosen, unrelated individuals of the same race. On the other hand, we also know that there are differences between populations of people that may be important for understanding differences in the frequency of certain diseases and the ways in which individuals metabolize drugs. These differences, however, are not fixed between populations and do not recapitulate historically constructed racial groups.

Pharmacogenomics is one area of medical science that is trying to sort through these differences. Variants in the enzyme CYP2D6 seem to correlate to the metabolism of at least 40 drugs, including beta-blockers, codeine, and some types of antidepressants. One study shows that between 5% and 10% of Europeans versus only 1% of Japanese, for example, do not produce CYP2D6 and cannot metabolize this group of drugs, resulting in dangerous side effects. (43) Yet even though we know genetic variation between populations exists, recent studies have shown that commonly used ethnic and racial categories do not accurately reflect human genetic variation. (44) If this is the case, can ancestry really be useful in accurately determining drug choice? Scientists know that such factors as sex, age, diet, and the environment can also affect the efficacy or potential reaction of a patient to a drug. Perhaps doctors will have to take these factors into consideration, too, when looking at possible pharmacogenomic outcomes.

In the end, the best pharmacogenomic information will come only from the individual's genome itself. Because of the potential shortcomings of population-based medicine, many believe that only by testing individuals for the genetic variations that cause adverse drug reactions or clinical ineffectiveness of a drug can pharmacogenomics be effective. We are not quite there yet, however. In the future scientists hope to chart a patient's drug profile and create a treatment precisely tailored for any given disease. This will occur only when the genes and/or SNPs involved in adverse or positive drug reactions and treatment failures or successes are identified and when the cost of testing for these locations is no longer prohibitive.

Pharmacogenomic research will present policy makers with many challenges. As pharmacogenomics changes the research and development process for drugs, the Food and Drug Administration will have to reconsider the way drugs are tested and marketed. Clinical trials are currently designed to study drugs produced for the average patient. What will happen when drugs are designed for an individual or a population group? Another area of concern for policy makers is the question of how pharmacogenomic data compare with genetic testing for various diseases. Unlike genetic tests, pharmacogenomic tests do not necessarily indicate the likelihood of an individual's developing a disease. Instead, pharmacogenomic tests can show how genetic variation can influence a patient's reaction to a drug. (45) But, as with genetic testing, concerns about the potential misuse of this information, particularly by insurance companies, remain. Pharmacogenomic testing could, for example, potentially correlate a negative drug response with "a more rapid progression of the disease or a worse outcome." (46) University of Pennsylvania bioethicist Arthur Caplan points out that "while genetic knowledge could revolutionize the way diagnosis is done in medicine, its ability to do so will not be quickly put into practice while patients are concerned that the direct result of a negative finding will be the loss of their insurance coverage." (47) The privacy of genomic information will remain a concern as long as no comprehensive anti-genetic discrimination legislation exists at the federal level.

YOUR GENES AND YOUR ENVIRONMENT

One of the dangers of the Human Genome Project is that it may suggest that genes are the cause of all disease and the primary or sole arbiter of human health. We cannot emphasize firmly enough that this is not the case. As we have already stated, only 5% of human disease is caused by a single gene acting alone without environmental influences. (48) Almost all human disease, including heart disease, diabetes, and most cancers, has instead a complex origin that is at once genetic and environmental. Scientists have long observed the association between environmental exposures and illness. We are aware, for example, of the dangers of smoking and its impact on human health. (49) Yet, according to recent studies, the horrible toll of tobacco smoking can vary considerably; smokers are susceptible to different illnesses depending on a variety of factors that include their genetic makeup. (50)

Understanding the interaction between genes and environment has been the work of the Environmental Genome Project, an effort under way since 1997 at the National Institute of Environmental Health Sciences (NIEHS). Using genomic tools, researchers at the project hope to identify genetic polymorphisms associated with individual responses to environmental exposures and to further study the environmental triggers involved in the development of human disease. (51) This information can be used to protect all species, both human and nonhuman, from environmental hazards. Microarrays, for example, can be used to monitor "the effect of potential contaminants on the gene-expression profiles" of different organisms. (52)

We already know that because of human genetic variation individuals have different susceptibilities to different environmental exposures. Much as pharmacogenomics is helping identify how human genetic variation can cause differential response to pharmaceutical products, the Environmental Genome Project is helping identify the ways in which human genetic variation can cause differential response to the toxins and carcinogens in our air, water, and diet. In other words, our genes can leave us susceptible to the environment in our own unique ways. The project has already begun to identify human "environmentally responsive genes" that seem to play an important part in human reaction to environmental exposures. (53) Remember, though, that susceptibility is no guarantee of outcome; it is only an assessment of risk. In April 2003, NIEHS Director Kenneth Olden announced the completion of the first phase of the Environmental Genome Project—a growing library of variation in 200 genes "involved in everything from processing toxins to fixing damaged DNA." (54)

Another important component of the project is focused on better identifying the toxins and carcinogens dangerous to humans. This can be done with the latest genomic technology, including high-throughput sequencing, microarrays, and bioinformatics. In the United States, the National Toxicology Program is charged with identifying dangerous environmental exposures. The program annually issues the "Report on Carcinogens," which identifies substances that are either known or anticipated to be human carcinogens. Federal and state governments depend on this information to help regulate both occupational and everyday exposure to substances like asbestos, benzene, environmental tobacco smoke, vinyl chloride—all known to be human carcinogens—and acrylamide, diesel exhaust particulates, and ultraviolet radiation A, B, and C—all reasonably thought to be human carcinogens.

(55) To determine dangerous exposure levels, scientists primarily use animal models complemented by epidemiological studies. The field of toxicogenomics, which applies genomic techniques to the study of toxicological effects, offers the possibility of more accurately identifying toxic and carcinogenic substances. Through the use of microarrays and other genomic technologies, toxicogenomics will help "survey the entire human genome and thus determine which genes are affected by specific chemicals." (56) It may also reduce the use of animals in studies of these dangerous substances. (57)

Everyone has genetic reactions to the environment. Toxicogenomics will be successful only when it is able to sort through the thousands of genes affected by environmental exposures and determine which gene expressions are potentially dangerous. Early data have been promising. One study has found that smokers who carry genetic variants in two particular genes have a 10-fold increased risk for developing bladder cancer. (58) Other studies indicate that specific occupational exposures may be amplified for people who have a particular genotype. (59)

Environmental genomics and toxicogenomics may prove both useful and controversial in the workplace. What if workers could be screened for susceptibilities to different environmental exposures or workplace hazards? If a genetic test can identify susceptibility, what kind of responsibility does an employer have for its employees' safety? Should there be government-issued warnings urging genetic tests before employment in certain jobs? Is it the job of employers, government, and industry to make the workplace safer or to exclude people from the workplace because of the results of a genetic test? What if someone is found to be susceptible to a particular exposure? Will this technology push companies to hire only workers without susceptibilities rather than enhancing working conditions?

Beryllium, a metal used primarily in the defense and aerospace industries, can cause a debilitating and sometimes fatal lung condition known as chronic beryllium disease, or CBD. For most, exposure to beryllium particles does not cause CBD; however, approximately 5% of beryllium workers contract CBD, sometimes more than 10 years after initial exposure. Scientists have discovered what seems to be a genetic risk factor for the disease—a single nucleotide polymorphism. (60) The hope of some is to identify workers with this genetic risk, allowing the workers to make "more informed medical decisions about whether they should work with beryllium." (61) With this information, at-risk workers could also be more closely monitored for early signs of the illness. Even though screening programs could protect those susceptible from working in such conditions, the current test for CBD is not necessarily predictive. In the case of beryllium susceptibility only a small percentage of those who have the genetic risk will develop the disease when exposed to the metal. But one doctor who treats CBD patients wonders whether such genetic susceptibilities will be used "as a way to blame the worker." (62) This type of testing, if abused, raises serious concerns about genetic discrimination in the workplace.

The improved identification of toxic and carcinogenic substances will surely play a beneficial role in our lives by preventing exposures to toxic chemicals and by enhancing our ability to identify individual susceptibility. Such prevention efforts may someday allow people to make decisions to avoid certain substances in their diet and their workplace. But these discoveries are also sure to challenge us in new ways. One fear is that environmental genomics "could inadvertently shift responsibility away from employers or governments onto individuals." (63) If people have testable sensitivities to known toxins and carcinogens, some might say, then maybe they should just avoid them. Although this may seem an unlikely result, it is possible that the genome project's emphasis on genes may favor genetic over environmental causes of illness. If this happens, personal genetic responsibility could gain favor over government regulation of toxic and carcinogenic substances, with terrible consequences for both the environment and for the health of all species on our planet.

FIXING OUR GENES

Genomics has also raised interest in attempts to physically repair the genes in our genomes to cure or even prevent many diseases. Gene therapy, fixing genetic malfunctions by repairing flaws in our genes, although still only a few decades old, may one day become a standard approach to treating disease. Thus far this therapy has very few successes to show, but most scientists believe that there is tremendous promise for its future. Studies are now under way exploring the possibility of using gene therapy to help treat a range of conditions including heart disease, cancers, AIDS and other immune disorders, muscular dystrophy, diabetes, and cystic fibrosis.

The science of gene therapy needs genes to find cures. So identifying the location and function of genes in the human genome will be a boon to gene therapy researchers. Theoretically, if a scientist knows the exact gene that contributes to the cause of a disease, then it should be possible to fix the genetic problem by either adding a working copy of the gene or repairing the gene. This simple idea, however, is very difficult to accomplish in practice. There are two types of gene therapy: Somatic cell gene therapy repairs or replaces a malfunctioning or missing gene in somatic or nonreproductive cells, and germline therapy does the same in germline or reproductive cells. Perhaps the most important difference between these two therapies is that the changes brought about by somatic cell gene therapy will not be passed on to offspring, whereas germline changes are passed on through reproduction. Such alterations to the germline, if possible with human cells, are sure to raise a host of moral and ethical concerns about the nature of our species and the limits of human technological power. In Chapter 4 we discuss the potential scientific and social impact of human germline gene therapy.

Successful somatic cell gene therapy is contingent upon finding ways to deliver a gene to its target. In gene therapy jargon the gene delivery mechanism is known as a vector. Two types of vectors are generally used for gene therapy: viral vectors, which take advantage of the normal behavior of viruses to deliver the gene, and nonviral vectors, which directly deliver DNA into a cell. (64) Some nonviral methods include technologies like microinjection, which uses micropipettes to deliver DNA directly into a cell, and electroporation, which uses electrical currents to coax a cell into taking in DNA. (65) Another nonviral vector technology is known as "naked DNA," which is just large amounts of a particular gene injected directly into targeted tissue. Cardiologist Jeffrey Isner and his colleagues at Tufts University School of Medicine developed a gene therapy that inserted naked DNA, a gene called *VEGF*, into a patient's heart muscle through a catheter. When successful, the genes take up residence in heart muscle cells and begin producing proteins that in turn help make new arteries. The technique has shown early promise. In a phase I trial, patients treated with the genetic injection developed new arteries around blockages. It is hoped that treatments like this one may someday be an alternative to invasive bypass operations. (66)

Despite such advances, viral vectors remain the most widely used vectors in gene therapy. Among viral vectors, retroviruses are currently the most popular method of gene delivery. A retroviral vector is made by removing the protein coding sequences from a retrovirus, such as Moloney murine leukemia virus, or MLV. Without these protein sequences the virus is essentially defanged; it can no longer infect someone with the disease it had once carried. The gene to be transferred, also known as a transgene, replaces the virus's proteins. The vector works by integrating the gene into the DNA of targeted cells. Unfortunately, gene placement is inexact and can occur anywhere in the cell's genetic material. For these types of integrating vectors, the transferred DNA will generally be sustained when a cell divides. Other types of viral vectors insert DNA only into the cell nucleus and not directly into its genetic material. These types of "nonintegrating vectors," as they are known, generally do not deliver DNA that is sustainable when a cell divides. (67)

A retroviral vector was used in a gene therapy protocol to treat the immune disorder human adenosine deaminase severe combined immunodeficiency disease, or ADA SCID. White blood cells were taken from a patient and mixed with a retrovirus engineered with a working copy of the human ADA gene. The "infection" of white blood cells by the retroviral vector delivered a working copy of the gene. Once the gene had been delivered successfully, the cells were reintroduced to the patient's circulatory system. Because these white blood cells have a limited lifespan, patients with ADA SCID must receive these infusions periodically to survive. Scientists are now developing new gene therapy techniques to cure ADA SCID so that ongoing gene infusions will not be needed. (68)

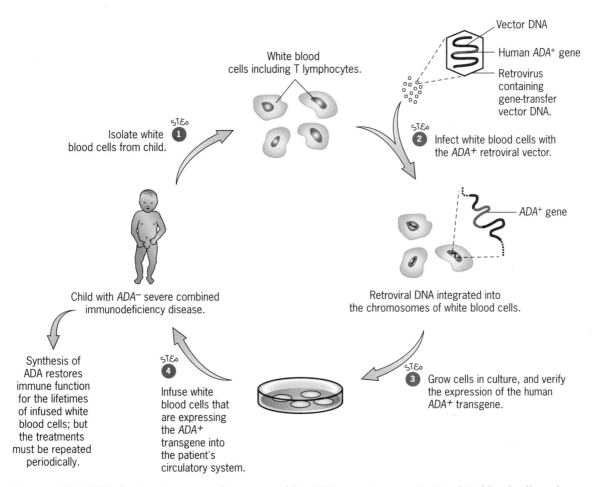

To treat ADA SCID doctors insert a working copy of the ADA gene into a patient's white blood cells and then infuse these cells back into the patient.

However, gene therapy comes with dangers, too. Because viruses are foreign bodies, there is a risk of an immunological reaction against viral vectors. In Chapter 4 we discussed the death of Jesse Gelsinger in a gene therapy protocol at the University of Pennsylvania Institute for Human Gene Therapy. Gelsinger was born with the rare metabolic disorder ornithine transcarbamoylase deficiency and died from a severe immune reaction. (69) Gelsinger's death, the first in a gene therapy trial, highlighted the dangers of vector-related immune response. Other potential side effects include the remote possibility that even though the virus's pathogenic material is removed before gene therapy, the viral vector could revert to its infectious state and possibly harm the patient. Another carefully watched side effect is cancer, which can occur because of the inexact and unpredictable nature of viral vectors. (70) If a viral vector delivers the transferred gene into the genome at or near a cancer-causing gene, the transfer can turn the gene on and result in cancer. Current technology limits our ability to deal with such an unwanted side effect. (71) These risks are taken into consideration in the development of gene therapy protocols. For many patients with an already deadly disease, the risks can be worth taking.

These risks can be avoided by developing vectors that place genes into a precisely chosen location in the genome or by directly repairing or replacing the malfunctioning or nonfunctioning gene. This type of genetic surgery is not yet feasible in humans, although there has been early success in the laboratory. For example, in 2002 scientists in Italy and in the United States replaced faulty DNA in cultured cystic fibrosis cell lines. (72) For now, however, somatic cell gene therapy technology allows only gene addition, not gene replacement.

The technology of somatic cell gene addition therapy has been slow to develop. For example, the exact genetic error that causes sickle-cell anemia has been known since 1957, when it became the first genetic disorder for which the underlying molecular basis was uncovered. (73) Still no genetic therapy for the disease has been developed, although progress is being made. In 2001 scientists used gene therapy to cure a form of sickle-cell disease in mice. (74) This success may some day be translated into success with humans.

Most of us can thank our immune systems for shielding us from common germs. People with X-linked SCID lack such defense. Until recently, children born with X-linked SCID survived only by living in sealed germ-free plastic enclosures or receiving bone marrow transplants. The mutation that causes this particular type of SCID lies on the X chromosome and occurs only in males. In France in 1999 doctors successfully corrected the faulty gene in stem cells of babies suffering from X-linked SCID. The gene therapy successfully spurred the immune systems of these SCID babies to develop and function properly. (75)

Despite the tremendous success in treating SCID, the gene therapy had a terrible side effect. Two patients developed leukemia because the working copy of the gene was inserted into or near a cancer-causing gene. The proximity of the placement activated the cancer-causing gene. Thankfully, the two boys are so far responding well to treatment. Trials of the X-linked SCID were suspended for almost two years pending review. Scientists are now searching for alternative vectors that can be effective and safer than the retroviral vector used to treat X-linked SCID patients. Gene therapy researchers are always seeking to improve upon current techniqes to repair and deliver genes, as well as to develop new technologies in this area. (76)

Gene therapy is a young science with a short track record. In fact, despite the successes of X-linked SCID gene therapy, the development of leukemia in two X-linked SCID patients demonstrates that this technology comes with great risks.

While bioethicists have come to a general agreement that somatic cell gene therapy should be considered part of standard medical practice and that it poses a set of challenges very different from germline gene therapy, there are still risks to be considered. Also of great concern is that as these technologies become more successful and widespread, medical advances may outpace our ethical understanding of what all this science means. And, finally, as the tragic case of Jesse Gelsinger confirms, the business of biology can have dangerous effects on safety (see Chapter 4). Government regulatory bodies such as the National Institutes of Health Recombinant DNA Advisory Committee (NIH-RAC) and the FDA must continue to work to ensure patient safety in conjunction with the development of gene therapy.

With the development and implementation of some of the technologies discussed in this chapter, genomics is off to a fast start. In the future, discoveries and new techniques will either improve or replace these technologies. Nonetheless, scientists have made an important beginning. We have yet to identify all human genes and characterize their function, both as single units and in conjunction with other genes and the environment. This task will consume researchers for many years to come as they continue to decode our genes and their functions and to seek out new ways to improve medicine and public health using our growing knowledge of the genome.

Despite this optimistic vision of the future, we still worry about the risks that these new technologies bring. Can our passionate desire to improve our health have its own unwanted side effects? Will we ironically put ourselves in danger in the process of finding cures and treating diseases? We can make these decisions only by both understanding the nature of these technologies and examining the ethical challenges that accompany them. To do otherwise would undermine the promise of the genomic revolution.

The World To Come: Agriculture

An August 2003 poll asked Americans if they had ever eaten genetically modified foods? Only twenty-four percent of the respondents answered, "Yes" (1) In fact, most Americans have knowingly or unknowingly eaten a genetically modified (GM) food, usually varieties of corn or soybeans. The United States planted 105.2 million acres of GM crops in 2003; another 167 million acres of genetically modified crops were planted around the world. Consumer groups estimate that approximately two-thirds of the processed foods sold in the United States contain ingredients from GM foods, also known as genetically modified organisms or GMOs. (2) But most of us have little information on GMOs. The Food and Drug Administration does not require labeling of most GM products, making it impossible for us to know whether we are purchasing and eating something genetically modified.

For thousands of years humans have domesticated their food sources. From cattle to corn, farmers have bred favored traits into their foodstuffs, creating new species of crops and animals to suit their needs. Modern corn, for example, is the product of more than 5000 years of selective breeding. Corn's most likely ancestor, a wild grass from Mexico known as teosinte, contains only a few edible kernels and looks little like the corn we eat today. (3)

Over many centuries, farmers across the Americas developed new varieties of corn, selecting for more kernels, higher crop yield, pest resistance, tolerance to specific growing conditions, milling quality, and flavor. All of this was done without any knowledge of genes, genetics, or genomes, as was all breeding until the application of Mendelian genetics. Even with a basic knowledge of genes, agriculture drew on little more than exceptional observational skills and a rudimentary understanding of heredity to breed more suitable varieties. But with dramatic advances in the past few decades biotechnology now dominates the once primitive

141

Welcome to the Genome, by Rob DeSalle and Michael Yudell.
ISBN: 0-471-45331-5 Copyright © 2005 Rob DeSalle and Michael Yudell.

processes of the farm. Agricultural breeding techniques have rapidly moved into the genomic age.

To many, the application of biotechnology to agriculture seems to be playing God with the very essence of life. The United Kingdom's Prince Charles, a fierce critic of GM foods, has said, "I happen to believe that this kind of genetic modification takes mankind into realms that belong to God, and to God alone." (4) By this reasoning, because we have long manipulated the genomes of our food sources we have been playing God with agriculture for millennia. "Every crop we eat today is genetically modified. Human beings have imposed selection on them all," says Susan McCough, a rice specialist at Cornell University. (5)

But is there something fundamentally different, something unique, about creating, let's say, insect resistance through the physical manipulation of genes versus selectively breeding that trait through traditional agricultural methods? What is the real meaning of inserting genes from one species into another's genome? To be able to begin answering these questions, we need to know just what goes into creating a GMO.

By definition, GMOs have a working copy of a gene from another variety of the same species or from a foreign species inserted into their genome. You have probably read about this type of modification: corn that has been modified with a gene from bacteria that produce a natural insecticide, soybeans engineered with a gene that confers resistance to a herbicide, and salmon engineered with an Arctic flounder gene, allowing them to mature faster in cold water.

Genetic modification techniques have advantages over traditional agricultural methods, the most obvious and important being the time it takes to cultivate the desired trait. For example, selecting for pest resistance or environmental tolerances took many generations of trial-and-error selective breeding. Through direct genetic manipulation the desired trait can be inserted into the genome after the genetic origins of that trait are identified.

How exactly is an agricultural product genetically modified? With plants there are three main ways to insert genes: the use of a bacterium to transport the gene; biolistics; and electroporation. Bacterial transfer of a gene is accomplished through

It took many millenia of breeding by peoples in the Americas to breed modern corn. The corn we know today was bred well before Europeans arrived in the Americas. Teosinte (left) is believed to be the ancestor of modern corn (top). An intermediary hybrid is also shown.

the use of the common soil bacterium *Agrobacterium tumefaciens*. This bacterium has the ability to transfer a segment of DNA (transfer DNA or T-DNA) into the nucleus of a foreign cell, thereby integrating novel DNA into the host's genome. Under natural conditions *Agrobacterium* infects plants at the site of a wound or graze, causing cells to turn into a gall or into tumor cells. These tumor cells are not related to human cancer or tumor growth. Beginning in the late 1970s, scientists developed a method to use the T-DNA in *Agrobacteria* as a delivery mechanism or vector for introducing genes of their own choosing into plants. The scientists removed the genes that caused the gall or tumor growth from an *Agrobacterium* and inserted the chosen gene in its place. (6)

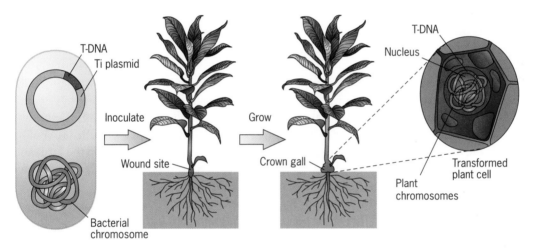

When engineered with a desired stretch of DNA, the common soil bacterium *Agrobacterium tumefaciens* has proven to be one of the most useful methods of gene transfer.

Today, the most common and crudest method of gene transfer is biolistics, or the use of a gene gun. This technique utilizes a modified .22 caliber gun that literally shoots plant tissue with tiny gold or tungsten particles coated with the desired T-DNA. After the gun is shot into a plant cell or directly into a plant leaf, some DNA will come loose from the metal particles and integrate into the host's DNA. This is a trial and error process that takes time, but once a successful shot is accomplished, plants with the new DNA can be bred for planting. (7)

In 1982, electroporation, the most versatile yet most difficult gene transfer technique, had its first success. The 1982 experiment used mouse cells. The procedure has since been shown to work with animal and plant cells as well as with fungi and bacteria. In this procedure electrical pulses are administered to cells, causing the pores of the cell membranes to open. Foreign DNA travels through the pores and takes up active residence in the cell, integrating into the host's genome. (8)

Using these techniques scientists and biotechnology companies joined forces to initiate an agricultural revolution. This union of agriculture and biotechnology, known in the business simply as ag-biotech, has rapidly and quietly introduced GMOs into agriculture. Remember our poll respondents, only twenty-four percent of whom believed that they had eaten a GMO? In the United States in 2003, 81 percent of the soybean crop, 40 percent of the corn crop, and 73 percent of the cotton crop are of the genetically modified variety. (9)

Why should we genetically modify plants? Are there compelling arguments in favor of the continued development and proliferation of ag-biotechnologies? Many advocates of these crops say that it's all in the numbers. By the year 2050 it is projected that the world's population will exceed 9 billion people, many of them living in developing countries. Advocates of GMOs assert that bioengineering can cut costs *and* boost the quality and quantity of food—improvements necessary to feed the world's burgeoning population. Advocates also believe that some GM crops can decrease the amount of pesticides used in farming, thereby limiting dangers to the environment. Meanwhile, critics worry about the potentially dangerous impact of these technologies on the environment and on human health. (10) Below we examine agricultural biotechnology products already common on the farm and look at the future uses of these technologies, highlighting some of the advantages and disadvantages of a genetically modified future.

BT CORN

Each year the European corn borer causes 1 billion dollars of damage to the United States corn crop. Once these pests burrow into corn stalks, insecticides are useless and the crop is lost. (11) A staple of both organic and nonorganic farming is the natural insecticide *Bacillus thuringiensis,* or Bt, a species of bacteria that produces a toxin that has long been considered a safe and effective killer of the corn borer. Commonly found in soil, Bt has proteins that kill the corn borer by fatally disrupting its digestive tract. Bt does not have the same effect on humans, and Bt corn is considered safe for human consumption. (12)

By placing a gene that expresses the Bt toxin into the corn genome with a gene gun, scientists were able to create transgenic Bt corn. This process began by first mapping the genes of *B. thuringiensis*. Once the gene that would express the needed toxin was found in *B. thuringiensis*, it was isolated, cloned, and transferred to the corn genome. (13) The new Bt seeds were collected and checked for the presence of Bt toxin, and classic genetic crosses were set up to produce a strain of corn that was homozygous for the engineered Bt toxin gene.

In 2003, it was estimated that 40% of the corn crop in the United States was genetically modified.

Critics of Bt corn, citing a study done by researchers at Cornell University, have argued that the presence of Bt pollen is doing great damage to the monarch butterfly population. The study suggests that monarch butterfly caterpillars are killed when the Bt corn pollen falls on the leaves that this species feeds on. (14) Field tests by another group of scientists offers contradictory evidence. Their study shows that there is no significant difference between butterfly caterpillar survival in fields planted with Bt variety versus non-GM corn. Richard Hellmich, a U.S. Department of Agriculture research entomologist at Iowa State University, believes that "if there are any differences out there, they aren't very profound." (15)

Another fear concerns the potential escape of Bt corn into the wild, where it could theoretically have a negative impact on its wild relatives. The Environmental Protection Agency requires buffer zones between GM and non-GM fields. In North America corn has no wild relatives, so gene flow between corn fields is not a concern, but in the rest of the Americas corn's wild relatives still thrive. Is there a real threat to native corn across great distances? The answer to this seems to be both yes and no. One study suggests that there has been gene flow between Bt corn and native maize in Mexico. Although scientists believe that this gene flow is of no threat to nontransgenic maize, it does demonstrate that this type of gene flow is possible. (16) Because plants can and often do hybridize between closely related species, it is possible that the Bt gene in modified corn can jump to other closely related plants and cause harm. Finally, there is the possibility that the widespread introduction of Bt corn will greatly reduce the use of conventional strains. This reduction in crop diversity can result in vulnerability to an unanticipated pest or disease, with a worst-case scenario of widespread destruction of the crop.

ROUNDUP READY SOYBEANS

One of the most widely used GMOs is the Roundup Ready soybean. This product, engineered by scientists at the biotechnology giant Monsanto, is part of the first generation of GM foods to be marketed to consumers in the United States. The Roundup Ready soybean was developed as a bioengineered complement to another Monsanto product, the herbicide Roundup. Roundup, or glyphosate, is considered to be a safe and effective herbicide. It works by destroying the weeds that may attack soybean plants by inhibiting the function of an enzyme called EPSP synthase. This enzyme produces three amino acids—tryptophan, tyrosine, and phenylalanine—that are all essential for protein synthesis in plants. Without these amino acids a plant will die. Roundup Ready soybeans are engineered to withstand the deadly effects of Roundup, allowing more soybean plants to survive herbicide spraying and therefore increasing crop yield. (17)

The Roundup Ready seed was developed by exposing petunia cells *in vitro* to glyphosate. Although almost all the petunia cells died after an initial exposure to the herbicide, some cells survived. These cells had a mutant, or alternate form, of the EPSP synthase gene that allowed the plant to survive in the presence of the herbicide. Once this gene was isolated, it was transferred, with some modifications, into the soybean genome by the *Agrobacterium* gene transfer method. The mutant petunia gene now provides soybeans with protection from Roundup. (18)

The use of Roundup Ready soybeans has resulted in increased spraying of the herbicide Roundup. This is because farmers can more liberally apply the herbicide to their fields without damage to the herbicide-resistant soybeans. Despite the increase in Roundup use, total herbicide spraying on Roundup Ready fields is down. This is probably beneficial to the environment because the herbicides displaced by Roundup are considered more damaging to the environment. (19)

RICE

Genetic engineering may offer the potential to create healthier, more nutritious foods such as golden rice, whose yellow kernels contain beta-carotene, a source of vitamin A.

Golden rice has its distinct color because of
the beta-carotene it is engineered to produce.
Advocates of the grain hope that it can help
supplement vitamin A intake in
undernourished areas of the world.

Each year, more than a million people die and a half-million go blind because
their diets lack vitamin A. Because rice is a staple food, particularly in poor and
developing countries, boosters of golden rice claim that it will both save lives
and increase the quality of life for many. To create golden rice, the beta-carotene
gene from a daffodil plant was inserted into the rice genome with *Agrobacterium*.
(20) Ingo Potrykus, the inventor of golden rice, hopes that farmers in developing
countries will use this technology to increase nutritional benefit for malnourished
populations. Some studies have estimated that children would have to eat up to
15 pounds of golden rice to satisfy the daily requirement for vitamin A. Contradictory studies suggest that normal portions of the rice provide sufficient amounts
of vitamin A. Even so, the farming of certain vegetables and the consumption of
conventional brown rice would avoid any need for a GM product, as both offer
greater nutritional benefits than golden rice. (21) But this criticism may fail to take
into consideration how dietary traditions might restrict such practices.

MORE REPORTS FROM THE FARM

Other foods are being genetically modified to increase nutritional content. Potatoes,
for example, are Americans' favorite vegetable and are, unless soaked in oil, a low-
fat, nutritious part of a diet. Unfortunately, most Americans eat their potatoes
deep-fried. The average American consumes over 140 pounds of potatoes every
year, primarily in the processed form of French fries and potato chips. Scientists
have developed a high-starch potato that absorbs less oil when it's fried. Because it
absorbs less oil this transgenic potato may save industry money on processing costs
and has the potential to make consumers a little healthier without changing their
dietary habits because of the reduced calories. (22) The results could be junk food
that's just a little bit healthier or a little less junky, depending on your perspective.
Experiments are also being conducted with potato plants engineered with a
gene from a jellyfish to produce a fluorescent protein when the soil becomes too dry.

Light-sensitive instruments can detect the faint glow of the activated fluorescent proteins and alert farmers that their fields need watering. A few sentinel plants in a field could prevent over-watering and also reduce the amount of fertilizer washed into water supplies. (23)

Coffee is another important part of the American diet. A welcome pick-me-up for most people, caffeinated coffee can cause anxiety, palpitations, high blood pressure, or insomnia in some. Approximately fifteen percent of the coffee consumed in the United States is decaffeinated. Unfortunately, the caffeine removal process reduces flavor, uses unhealthy chemical solvents, and still leaves some caffeine in the beans. At Japan's Nara Institute of Science and Technology scientists have found a way to genetically decaffeinate coffee beans, reducing caffeine by up to 70%. In the future scientists hope to increase that percentage. For now, though, the beans will remain off the shelves. Starbucks says that it doesn't want GM coffee beans, Hawaiian coffee growers have demanded a moratorium on GM coffee, and the European Union says it will not import the product. (24)

Saturated fats increase the risk of clogged arteries and heart disease. So oil-producing seeds like soy, corn, and canola are being genetically modified to contain less saturated fat. They are also being altered to contain more vitamins and to withstand higher temperatures, so they can replace less healthy animal fats in cooking. However, if these improved oils still contain the same amount of total fat, they will continue to contribute to obesity and other health problems. (25)

Plant biotechnology may also have important medical applications through the use of plants to deliver both drugs and vaccines. Pharmaceutical companies hope that in the near future farms will replace factories for production of some drugs. Already in the works are potatoes that are genetically engineered to contain a drug to treat cirrhosis of the liver, rice to treat cystic fibrosis, and tomatoes that produce an antihypertension drug. Research has also been conducted using bananas genetically modified to produce certain vaccines. Fruits that can be eaten raw like bananas are considered ideal for delivering vaccines because cooking the fruit would destroy the vaccine. The fruit could be puréed, packaged in small jars, and tested for dosage. Potato-based vaccines for hepatitis B and the Norwalk stomach virus are currently in an advanced stage of development, as are vaccines against cholera (tobacco), rabies (spinach), malaria (tobacco), and HIV (tobacco and black-eyed bean). (26)

Supporters of this technology note that producing pharmaceuticals in transgenic plants would be much cheaper than current methods. Plants cannot carry human diseases, so the drugs may be purer and safer than those derived from animals. Some drugs could potentially be stored and transported in dried seeds or fruit. Critics worry that pharmaceutical plants have the potential to release large quantities of drugs into the environment through the food and feed supplies or through accumulations in the water supply, soil, and other plants. (27) It is also theoretically possible that drugs may even end up in the human food supply—either from accidental mixing of plant products or from pollen drifting into nearby fields and cross-breeding with other crops. This could increase the percentage of bacterial resistance to antibiotics and other drugs.

A GENETICALLY MODIFIED ZOO

Plants are not the only agricultural products that are genetically modified. Animals have also been bred for traits using both traditional agricultural methods and bioengineering. Genetically engineered salmon, for instance, can reach a market size of 7 pounds in 18 instead of the normal 36 months. By inserting genetic material from two other fish species into Atlantic salmon eggs, scientists stimulate the salmon's growth hormone, enabling the fish to grow year round, even in colder waters. Researchers say they can also modify a dozen or more other species of fish, potentially leading to a "blue revolution" in the fisheries industry. (28)

These 18-month-old salmon have been bred very differently—the larger GM salmon has been engineered to grow all year round and has already reached market size.

Modified carp, tilapia, and other fish might ultimately provide savings to consumers as well, especially people in developing countries, where more affordable protein sources are badly needed. Although consumption of GM fish presents no known human health risks, there are serious concerns that these fish, if they escape from fisheries into the wild, may have an unintended and unpredictable impact on wild salmon species and on the overall ecosystem. (29)

Health-conscious consumers want less fat in their pork, but when a farmer breeds his leanest pig, only half its genes are passed on to its offspring and only one-quarter reach the following generation. Because cloning creates identical twins of the original, farmers are looking into cloning pigs and other animals, including chickens that lay eggs low in cholesterol. (30) Cloning animals concerns some critics, in part because genetic diversity within a species is very important for fending off disease. Observers are also worried about the overall genetic health of cloned animals. Recent studies have confirmed that with existing technology cloned animals exhibit a greater number of genetic defects than animals bred naturally. (31)

In 1996 Dolly the sheep became the first mammal cloned from an adult cell. Dolly was euthanized in 2003 after suffering from a degenerative lung condition.

Until recently one of the least-discussed areas of agricultural biotechnology was the use of animals to safely produce new classes of drugs and to provide additional sources for human organ transplants. Every year more than 6000 people in the United States alone die awaiting organ transplants. (32) Some scientists believe that the xenotransplantation (the transplanting of organs between species) of hearts, kidneys, and livers from pigs and other species into humans can solve the organ shortage. To prevent rejection of pig organ transplants, researchers have altered the genes in pig DNA that normally prompt the human immune system to recognize these organs as foreign. One concern is that pig DNA may contain viruses that could spread to humans. (33) Animal rights activists who oppose xenotransplantation maintain that programs promoting human organ donation would be much cheaper and safer and would spare the lives of animals. (34)

Genetically modified animals can produce a wide variety of substances with pharmaceutical applications. For example, goats are now being engineered to treat patients who lack antithrombin, an important anticoagulant and anti-inflammatory human protein. Approximately 1 in 5000 Americans are deficient in the protein and are at risk of developing thrombosis and other clotting-related disorders. By inserting a human gene into goat DNA, researchers have cloned goats whose milk produces antithrombin III, a valuable anticlotting drug. A similar process could create cows and sheep that produce other drugs in their milk, including treatments for cystic fibrosis, Crohn disease, and rheumatoid arthritis. (35)

Another interesting GM product concerns teeth. Sugar does not directly rot your teeth; the damage is done by the bacteria in the mouth and on teeth that come in contact with sugar. The human immune system fights bacteria by creating antibodies. Now scientists have genetically engineered tobacco plants to produce antibodies against *Streptococcus mutans,* the bacterial culprit of most tooth decay. When applied to people's teeth in clinical trials, this vaccine protected people from cavities for several months. (36)

MICROBES

Bioengineered microbes can be applied to industrial uses, as well as to medicine. In 1980 *Pseudomonas,* bacteria engineered to digest crude oil and clean up oil spills, became the first life-form and the first GMO to receive a patent. In a landmark ruling, the United States Supreme Court held that the engineered bacterium was a product of a person's work, not a product of nature, and was thus patentable. This ruling opened up opportunities for the patenting of genes and other genetically engineered organisms. (37)

Other organisms have been engineered to clean up the environment. The insertion of two bacterial genes into thale cress (*Arabidopsis thaliana*) creates a plant that can both tolerate arsenic-contaminated soil and absorb its toxin, storing it in its leaves. The engineered genes, which come from *E. coli* bacteria, produce enzymes that digest arsenic compounds. (38) One of the more fantastic applications of this technology involves the bacterium *Deinococcus radiodurans*. The Guinness Book of World Records lists *Deinococcus radiodurans* as the world's toughest bacterial species.

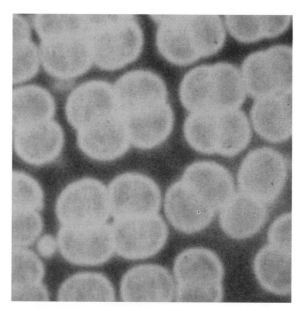

Although some of the characteristics of this species of bacteria might seem like something out of science fiction, scientists hope that *Deinococcus'* natural ability to withstand radiation can be utilized to clean up certain types of environmental hazards.

Deinococcus can survive radiation levels 3000 times the lethal dose for humans because it has a remarkably efficient DNA repair system. When exposed to high levels of radiation, the *Deinococcus* chromosome and DNA are blown apart, resulting in 100–200 breaks in the double helix. If placed into an aqueous environment after this exposure, the DNA and chromosome are repaired, and the organism will begin replicating again. Such high levels of radiation in organisms lacking this type of super-repair system would cause severe mutations, destroying the organism. Because *Deinococcus* can repair most of the genetic damage caused by extreme radiation, researchers hope to harness this bacterium to clean up nuclear waste sites. (39)

The environmental legacy of the Cold War is still being calculated. The nuclear weapons manufacturing program in the United States alone created a huge toxic waste problem. About one-third of the 3000 military waste sites contain dangerous organic chemicals and toxic metals mixed with radioactive waste. At many of these sites, waste is leaking from underground storage tanks and seeping into soil and groundwater. Over 1.7 trillion liters of groundwater is already contaminated, and the problem is spreading. (40)

There are genes in bacteria such as *E.coli* that can break down organic chemicals and stabilize or neutralize toxic metals, but at nuclear waste sites radiation levels would quickly kill these beneficial bacteria. By transferring key genes from *E.coli* and from *Pseudomonas* into *Deinococcus radiodurans*, researchers have discovered that these modified bacteria can help clean up these sites by breaking down harmful chemicals by, for example, converting the toxic metal mercury into a safer form. It is hoped that other strains of this bacterium can be engineered to break down cadmium and to transform water-soluble uranium into a solid that will not leak out of waste sites. Such modified bacteria may someday be available, but extensive safety testing will be required before they can be released into the environment. (41)

Land mines are another legacy of warfare. Every year 25,000 people are killed or injured by land mines, which are difficult and dangerous to find. Most mines leak trace amounts of explosives, making it possible to biologically sniff them out. To do this, researchers have modified *Pseudomonas putida* to glow in the presence of TNT. In field tests this bacterium revealed simulated mines under ultraviolet light. A variant of harmless soil microorganisms, the bacterium could be sprayed by hand or from crop dusters and would die in about a week. But before dying the bacteria could alert mine sweepers to where the danger is. (42)

ISSUES AND ETHICS
The Food Chain and Unintended Effects

It is feared that GM crops can affect the environment in unintended ways. The potentially unpredictable nature of the interaction between GM agricultural products and natural ecosystems has many worried about the consequences of ag-biotech. It is not that genetics, per se, is unpredictable. After all, genetics and the rules governing inheritance and population dynamics are highly predictable. It is the interaction with a heterogenous and uncontrolled environment that makes GMOs unpredictable. Some fear that pest-resistant corn might poison caterpillars and butterflies; animals that feed on them could be affected as well. Other indirect effects of GMOs could include a decline in songbird populations. This occurs because of the indirect effect of herbicide-tolerant crops, which let farmers virtually eliminate weeds from their fields. Because weed seeds are an important food source for skylarks and other birds, it is possible, according to some estimates, that in some areas the food available to these birds could be reduced dramatically. (43) Another unintended effect might occur if plants that are modified to produce pharmaceuticals are not properly handled. In such a case, these substances would enter the ecosystem and the food chain. What unintended effects would occur if wildlife like birds, insects, mice, and worms eat plants that contain drugs and vaccines or if soil or stomach bacteria encounter these drugs? We do not know. Finally, genes from modified plants can be transferred into the genomes of closely related species or hybridized with weeds, raising concerns that such new species could have a competitive advantage in the wild. Scientists also raise the possibility that engineered genes could, in theory, "enter the wider environment" and possibly be transferred into a new genome. The EPA has prohibited the planting of Bt cotton in parts of Arizona, southern Florida, and Hawaii, where it could possibly cross-breed with wild cotton. Farmers are also working on techniques to control potential gene flow by "segregating transgenic crops, planting 'buffer crops' to catch drifting pollen, and controlling the timing of flowering and seed production so that they are less likely to breed with either other crops or wild relatives." (44)

The benefits of agricultural biotechnology are clear and measurable, as are many of the risks. Some of the risks, however, are simply unknown and uncertain. (45) An assessment of these unintended and unknown effects may be prohibitively expensive. The challenge for policy makers is to make sure that existing regulations minimize these risks, that new regulations are put in place as needed, and, if necessary, that safety standards be designed that prohibit the development of certain types of GMOs where the risks are too great.

Will They Escape?

If modified fish burst out of their pens and into the ocean or pollen from GMO crops blows into a neighboring field, critics say the results will be unpredictable. The modified organisms might be hardier than native species, out-competing them and possibly causing extirpation events in the native plant populations. Other research indicates that modified fish might be good breeders but are not well suited to survive in the wild. If these modified fish were to breed with wild species and transfer weaker traits, it could theoretically cause a population crash.

Advocates of genetic engineering argue that modified plants and animals can, in fact, be successfully contained, but in real-world situations it does not always work out this way. Thousands of farm-raised salmon, for example, have already escaped from their pens with an unknown environmental impact. Only by sterilizing these modified organisms can we make sure that they can't have a genetic impact on the environment. Sterilization programs, however, are not completely reliable. Furthermore, if GM plants are rendered sterile, then birds and insects that normally depend on their seeds, nectar, and pollen will be deprived of food sources.

Reaching the Right Audience

Genetically modifying plants and animals can produce great benefit by reducing the cost of pharmaceuticals, foods, and other crops. Vaccines distributed to developing countries as food will help eliminate some of the barriers to lifesaving medicines. GMOs have the potential to vastly increase crop yields and improve nutrition across the globe. There are those who worry, however, that GMOs will, in the end, have little lasting benefit to the poor. (46) The high cost of developing and testing new products means, for now, that GM crops will cost more than conventional ones. Manufacturers may ignore the needs of the poor by focusing their efforts on products that make the highest profit. Expensive GM medical and agricultural products may, in a worst-case scenario, increase the gap in medical care and nutrition between rich and poor nations.

Big Business and Genetic Ownership

Many who have worried about the intentions of large corporate biotechnology companies have criticized the Roundup Ready family of engineered seeds, noting that Monsanto may be exercising too much control over farmers by creating what some see as a closed system: These Monsanto products must be used in tandem with one another to be effective, making the farmer dependent on the company for their use. Despite Monsanto's stated desire to use biotechnological tools to improve agriculture, nutrition, and human health, critics recall that Monsanto lost the public trust long ago through the development of the chemical defoliant Agent Orange, as well as the environmental contaminant polychlorinated biphenyls (PCBs). (47) Why should we trust Monsanto now, particularly with our food?

In December 1999 Monsanto, facing widespread criticism, announced that it was abandoning development of germinator control (terminator gene) technology—technology that genetically engineers crops to destroy their own seeds, requiring farmers to continually purchase new seeds. Scientists and executives at Monsanto believed that this technology would help protect their intellectual property rights and provide some protection against the environmental assimilation of GMOs because sterile seeds would theoretically be unable to reproduce and infiltrate the environment. (48)

Opposition to terminator technology was almost universal. A disparate group of critics, from Indian peasant farmers to Greenpeace to the president of the Rockefeller Foundation, criticized the technology as a threat to sustainable farming around the globe; they questioned its safety for human consumption and worried about its impact on worldwide biodiversity. In 1998, for example, India prohibited the import of terminator seeds, fearing that they would destroy the livelihoods of poor farmers. Critics also feared that subsistence farmers, who "often buy small amounts of improved varieties and breed them with local varieties to bolster yields," would be unable to afford the new seeds. (49) Finally, environmental critics were concerned that the pollen from the terminator-planted fields would spread to other plants, and cross-breeding (or hybridization) would put those plants at risk of being sterilized. (50)

Surprisingly, the move to suspend the terminator project came even though Monsanto did not, and never would, own the technology. Nor was the seed beyond the research and development stage. In fact, the terminator gene never made it out of the lab. (51) The uproar over an untested, unproven, and unused technology embodied the fear that corporate greed would drive Monsanto and like-minded biotechnology companies to devastate the environment, control worldwide agricultural communities, and introduce potentially lethal compounds into staple foods—all in the name of profit. Consequently, the terminator crisis intensified the fear of biotechnology and GM food that had already spread across Europe and has since spread rapidly around the globe.

Are Genetically Modified Foods Safe to Eat?

The Food and Drug Administration's GM food policy, based on existing food law (the Federal Food, Drug, and Cosmetic Act) "requires that genetically engineered foods meet the same rigorous safety standards as is required of all other foods." If GMOs *do not* contain "substances that are significantly different from those already in the diet," they *do not* require premarket approval. This means that most GM foods do not need FDA approval for sale in the marketplace because most GM foods do not contain significantly different substances (proteins, toxins, antioxidants, etc.). (52) The Act places the onus and legal duty on "developers to ensure that the foods they present to consumers are safe and comply with all legal requirements." However, premarket approval *is* required if the protein produced by bioengineered genes "differs substantially in structure and function from the many proteins that comprise our foods." (53) This does not mean, however, that GM products are unregulated in the United States. The FDA, the Environmental Protection Agency (EPA), and the U.S. Department of Agriculture (USDA) monitor GMO safety from product inception to market under a regulatory umbrella known as the coordinated framework. The coordinated framework for GMO regulation will be tested in the years to come as new types of genetically engineered agricultural products are brought to market. In January 2004, USDA officials announced that they would review and begin to revise regulations of genetically modified crops in the hope of meeting current and anticipated challenges. (54)

U.S. regulations have not always been a success. The fear of a GM food crisis in the United States seemed real enough in September 2000 when taco shells sold by Taco Bell were found to contain GM corn approved for animal, but not human, consumption. This strain of GM corn, produced by Aventis and known by the name Starlink, contains a protein that, unlike its GM counterparts approved for human use, may cause allergic reactions. The mixture of Starlink into other corn varieties touched off a firestorm of criticism of FDA and EPA regulations on GMOs and has forced both agencies to review their GMO policies. The EPA has said, in the wake of the Starlink situation, that it is highly unlikely that any GMO grains will ever again be approved for use solely as animal feed. (55)

Starlink brought home the potential reality of allergens and GMOS. Although allergens are a potential danger in all foods, their use in GMOs is troubling simply for the reason that after a genetic modification, foods that formerly did not present allergenic issues can become dangerous to some. The threat of an allergic reaction could be hidden from consumers if they were unaware of a genetic modification that incorporated a particular allergen. Soybeans modified with genes from Brazil nuts, a known allergen, caused allergic reactions in tests and had to be kept off the market. But because most food allergens come from known sources, primarily milk, eggs, wheat, fish, nuts, soybeans, and shellfish, scientists must take care not to incorporate the known allergenic protein into a new food source. (56)

Although GMO foods are probably no more a threat to human health than other foods, consumers still worry about food safety. It is worth pointing out as a point of comparison that conventional, nonbiotech, breeders sometimes create new varieties. Another way to create a new variety of plant is to expose its seeds to radiation and chemicals that induce genetic mutations. Unlike the controlled changes in DNA produced by genetic engineering, these mutations are random. Nevertheless, unease and a sense of not knowing the effect of GM foods on human health have led some manufacturers to stop putting GM ingredients into baby food. (57)

Should These Foods be Labeled?

In Japan, Australia, and countries throughout Europe, the labeling of GM foods is required. In stark contrast stands the U.S., which requires labeling only if a GM food product is "significantly" different from an unmodified one. (58) Supporters of GMOs say labels would simply scare people away from new products that are benefiting farmers, consumers, and the environment. Opponents argue that labels are needed because of the unknown environmental and public health threats of GMOs and that consumers should also have the right to choose whether or not to partake in this technology.

Labeling GMOs is a complicated process. GM products would have to be kept separate from unmodified foods at every stage of growth and preparation: separate fields, separate trucks, and separate grain silos. Tracking GMO products from seed to supermarket would require costly and time-consuming paperwork. Current segregation methods have failed on numerous occasions to keep GMO ingredients out of non-GMO foods. In 2000, the accidental mixture of Starlink corn seeds with other corn seeds confirmed this fear. (59)

Deciding which foods to label would be no simple matter. Highly processed ingredients like sugar, corn syrup, and vegetable oil apparently contain almost no genetic material or proteins from the plants they come from. But in Europe even refined oils must be labeled if they contain 0.9% GMO material. (60) A labeling program will require policy makers to decide where to draw the line regarding soft drinks that contain GMO corn syrup; pork from pigs that eat modified soybeans; and cheese, bread, beer, and yogurt made with enzymes from modified bacteria.

The most likely outcome of the labeling debate is that a niche market for GM-free food will be created. Perhaps the best model is the kosher foods market. Rabbinic authorities strictly supervise the production of kosher food, and rigorous procedures are followed in its preparation. Kosher foods serve a small group, but they are essential to that group and are guaranteed a small but permanent market. For those who are opposed to GMOs, the same type of niche market may be emerging, as evidenced by the number of organic food stores declaring themselves GM-free. (61) This approach may offer a compromise to government-mandated labeling.

From bananas that deliver vaccines to animals that carry organs for human transplantation to corn engineered to ward off its worst pest, the integration of genomic technology into agriculture offers many possibilities. This chapter has highlighted agricultural technologies that are changing our world—both the ways in which they can and will improve our lives and the ways in which they may be dangerous. The public discussions on this subject have for the most part been muddled. Critics of GMOs assert that there has been inadequate long-term testing of these products, that agricultural biotechnology is not an extension of traditional breeding methods, and that GMOs are safe neither for the environment nor for human consumption. At its extreme, critics mock agricultural biotechnology as "frankenfood", trying to convince the public that science has gone terribly awry. On the other hand, many in both the scientific and biotechnology communities claim that GM foods are as safe as non-GM foods. Is there a position between these two opposing views? Can we feel confident that there are at best only limited threats to human health (that need to be studied) and yet acknowledge that concerns about environmental safety need to be dealt with more judiciously? Without such a middle ground, it is possible that important GMO safety issues might get lost. It is also possible that the obvious benefits of the technology could be tossed aside.

At the beginning of this chapter we asked whether traditional agricultural methods significantly differ from agricultural biotechnology. Examination of the techniques and technologies of genetic modification reveal that there may be several answers to that question. Safety issues, as they arise, are for all of us to consider. To be responsible, we must promote public awareness and knowledge of these issues and, through discussion, research, and democratic processes, develop a consensus on what is best for both society and the environment. That will be the only way to ensure our health and safety and to protect the best of the genomic revolution.

Conclusion

Caution: Welcoming the Genome

Albert Einstein's work on relativity and quantum theory led to what was perhaps the greatest scientific revolution of the twentieth century. His contributions not only altered a basic understanding of the physical laws of the universe but also had a considerable impact on humanity's view of itself. Over the course of the century philosophers, artists, and writers considered the "remarkable and sometimes quite unforeseen cultural transformations and resonances" brought about by Einstein's ideas. (1) Einstein's mark was also felt in politics. As a refugee from Nazi Germany, Einstein was concerned with what the Germans could do if they harnessed the atom to promulgate their evil. Fearing such catastrophic repercussions, Einstein signed a letter to President Franklin D. Roosevelt in 1939 encouraging the United States to develop its own atomic device—a task made possible in part by Einstein's own theories. (2)

At war's end, with the Axis powers defeated, Einstein's view of the bomb changed and he is quoted as saying, "I think I have made one mistake in my life, to have signed that letter [to FDR promoting the atomic bomb]." Einstein also called on scientists "to do all in our power to prevent these weapons from being used for the brutal purpose for which they were invented." (3) Einstein's own scientific work and political mettle had had an impact on the development of atomic weapons. But Einstein didn't develop his theories of energy and matter for use in such ways. Instead, the application of these scientific ideas was a consequence of the scientific and technological development in which Einstein played an important role.

Although genomics, which is largely a collaborative enterprise, lacks the singular influence and charisma of an Einstein, this new science will surely have important and lasting scientific, medical, cultural, and political impacts. As Einstein once helped uncover the physical laws of the universe, genomic scientists are uncovering the biological mysteries of all species on Earth. In this book we have written about the ways in which genomics will some day improve health care and how genomic technology is changing the nature of agriculture. We have also written about the ways in which genomics is affecting how we understand ourselves as individuals and as a species and how humans and all other species on Earth are related to one another through their genomes. And, finally, we have explored some of the ethical and moral challenges we will face as the genomic revolution takes hold. All of these influences are sure to inspire philosophers, artists, and writers to think about the human condition in new ways. From our current vantage point, genomics will indeed be the greatest scientific revolution of the twenty-first century.

Although science itself may not have a political or social agenda, people often do and, if given a chance, can appropriate science for their own purposes. We know how eugenicists manipulated the meaning of human heredity to justify awful acts during the first half of the twentieth century. And we know from the Jesse Gelsinger case how conflicts of interest in biotechnology could cause great harm. Moreover, sometimes technologies are developed that can themselves be dangerous either in their original incarnation or in some permutation thereof. If we understand this, we can then work to mitigate or even deter what New York University sociologist Troy Duster calls the "social side effects" of genomic technology. (4) We cannot now know what all of these side effects will be, or how they will affect us, but we can recognize that both the science and the side effects are part of the same package. Genomics, like almost all technology, has the potential to live up to its billing as a marvel of human ingenuity or to disillusion us by causing great injury and harm. All stakeholders in the genomic revolution, from government regulators to scientists to citizens, must recognize the potential for damaging side effects and find ways, through the development of legislation, ethical guidelines, and cultural standards, to make sure that our genetic privacy is ensured, that genetic discrimination never comes to pass, and that eugenics remains an historical case study.

Unfortunately, in some cases, side effects have already been seen. But perhaps we can learn from our early failures. One of the most familiar genetic technologies is DNA fingerprinting, which is used primarily to help law enforcement solve crimes, to resolve issues of paternity, and to assist in postmortem identification. From television shows like *C.S.I.* to real-life courtroom dramas, most Americans are familiar with the basics of DNA fingerprinting. Despite the incredible commonalities in our DNA, all people, except identical twins, have unique genomes. Scientists can use the DNA taken from blood, a hair follicle, saliva, or even skin cells from a fingerprint to highlight unique portions of an individual's genome. (5)

In police forensics, the DNA fingerprint from a tissue sample can be compared with DNA taken from a crime scene, be it blood left behind at the scene or semen collected in cases of rape. In cases of paternity, DNA fingerprints are compared between a child and his or her suspected father. Over 342,000 paternity tests were conducted in the United States alone in 2002. (6) Finally, in cases of postmortem identification, DNA samples can be used to help identify the remains of individuals

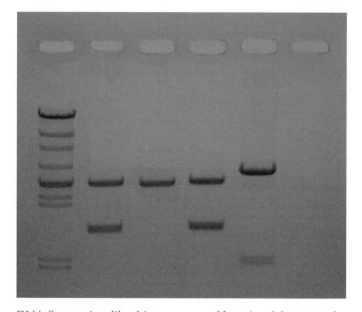

DNA fingerprints like this one are used by crime labs across the country to help identify suspects in crimes. This particular DNA fingerprint shows a test conducted on the same fragment of DNA from a crime scene sample compared to samples from three suspects. The second row or lane in the image is from the crime scene. Lanes 3, 4, and 5 are from suspects. The sample in lane 4 matches the crime scene. Identifying a person using this technology takes the agreement of several markers like the one shown here.

lost in war and other tragedies. Many of the victims of the World Trade Center terrorist attack have been identified through DNA fingerprinting. Victims of the atrocities in the former Yugoslavia are also being identified in this way. (7)

DNA fingerprints are incredibly accurate. Scientists estimate that if 10 DNA stretches from a genome are compared, there is only a one-in-a-billion chance of a random match between two people. (8) This pinpoint accuracy, attainable only when qualified scientists conduct testing, is enhancing law enforcement's tools to help identify and convict criminals. But mistakes can happen, and the side effects of this technology have quickly revealed themselves. For example, a man in Great Britain with an advanced case of Parkinson disease that severely limited his mobility, was arrested and charged with a burglary that took place over 200 miles from his home, based on a match in a DNA database. (9) A more precise test ultimately vindicated the man. Some errors, however, are not always accidental and are not nearly as simple to fix.

In Texas, the Houston Police Department closed its DNA crime lab in 2002 after an independent audit uncovered egregious problems that amazingly included the fact that none of the analysts who worked in the lab "were qualified by education and training to do their jobs." (10) Problems at the lab were compounded by substandard scientific methodology and a leaking roof. As of September 2003, 49 cases have been retested by a private lab, already leading to the discovery of 14 significant errors. (11) In one case, Josiah Sutton, a Houston teenager imprisoned on rape charges, was released after DNA evidence from his case was retested and showed him to be innocent of the crime. (12) Similar problems have occurred in other police departments across the country. Ironically, cases like Sutton's highlight how DNA fingerprinting, when used properly, can also vindicate once-convicted criminals. The Innocence Project, a nonprofit legal clinic run by lawyers Barry Scheck and Peter Neufeld, has used postconviction DNA testing to earn the release and exoneration of almost 100 individuals. As of April 2004, 143 inmates nationwide, including the cases of the Innocence Project, have been freed after postconviction DNA tests proved their innocence. Of those freed, 13 had been on death row. (13)

DNA use in forensics raises a host of other issues. In the United States all 50 states and the District of Columbia now have varying laws requiring the collection of DNA from individuals convicted of certain classes of felonies. (14) These samples are used to help solve crimes and are a part of the Combined DNA Index System, a coordinated database of local, state, and national DNA samples. Law enforcement investigators can electronically compare DNA profiles in the system to samples collected at crime scenes to look for matches. (15) Such DNA databasing makes privacy advocates nervous. How far will such databanking go? Will all arrestees be forced to provide DNA samples? Will these samples be destroyed if there is no indictment or conviction? What types of safeguards need to be in place to ensure that DNA information remains private and is used only for criminological work? And what types of uses of DNA information beyond crime scene matches should be allowed? Should a "law enforcement purpose" include, for example, acquiring information about a suspect's health status? And what about the power of DNA databasing to stigmatize both individuals and populations? A legal observer notes that the immutability of DNA "creates the spectre of prejudice. It allows an individual to be placed into a discrete class, cut along lines defined by the most intimate and private facts. . . . Seizure of such information, therefore, potentially gives the possessor the power to stigmatize and discriminate against many subjects." (16) Some have even suggested that the universal collection of DNA samples at birth might be a deterrent to criminal activity. (17) This "Big Brother" scenario would certainly face fierce opposition.

Not all genomic technologies will trigger such complex and potentially dangerous side effects. But as you can see, a simple genomic tool with what many considered fairly straightforward uses is turning out to be quite complicated in both its application and social impact. And the future seems no less simple. Genomics is sure to not only challenge our greatest scientific minds to continue to uncover the secrets of life but also to put to the test government and social institutions as they try to safely incorporate genomic technologies into health care.

There are also sure to be cases in which scientists develop technologies with applications that are morally problematic or unacceptable to many or all in our society. The idea of cloning humans has elicited this type of response from across the professional and political spectrum. Many scientists, clergy members, bioethicists, and laypeople have come out against the cloning of humans for reproductive purposes. One theologian called human cloning a "moral disorder." (18) In February 2003 the United States House of Representatives passed a ban on both reproductive and therapeutic cloning. (19) The Senate has not, as of April 2004, taken up this bill. Nine states, including California, Rhode Island, Virginia, and North Dakota ban human reproductive cloning. Cloning bans in Michigan and Iowa extend to cloning used for research purposes. (20)

Although most agree that human reproductive cloning is outside our moral boundaries, there is a diversity of opinion on non-reproductive forms of cloning, which generally fall under the rubric of stem cell research. This area of research is also known as therapeutic cloning, distinguished from reproductive cloning because its use cannot lead to a human being.

Stem cells are cells that are active throughout the life of an organism in making new cells for organs and other tissue. Stem cells in bone marrow, for example, are constantly making new blood cells. A stem cell is characterized by its ability to repeatedly divide, making exact copies of itself, and by its ability to differentiate into specialized cells found in all body tissue. These characteristics make stem cell research a highly promising area of study. Scientists are now experimenting with stem cells in healing spinal cord injuries, in replacing skin tissue, in organ repair, and in treating a wide range of disease including cancers and heart disease. (21)

But because stem cell research often uses stem cells taken from human embryos, their use is controversial. (22) By integrating the nuclear genome of an individual into an embryonic stem cell, that stem cell can then be copied and used in research and possibly some day clinical practice, in treating some of the disorders mentioned above. The transplantation of one individual's nuclear genome into an embryonic stem cell means that that stem cell's DNA will now match the donor's DNA, thus avoiding possible tissue rejection when the stem cells or products of the stem cells are used in the donor's body.

This research is controversial to some because embryonic stem cells are taken from unused embryos left over from *in vitro* fertilization techniques. In 2001, President Bush limited the use of embryonic stem cells in federally funded research to an existing sixty cultured lines. Many scientists and public figures opposed this move, questioning both the long-term viability of these stem cell lines and whether there were actually so many. It is possible that the President's decision to limit federally funded stem cell research will propel this avenue of scientific investigation into the private biotechnology sector, where research would be conducted outside of federal ethics guidelines. None the less, research on stem cells continues. Recent advances in the uses of adult stem cells may someday provide additional avenues of research in this area. (23)

President George W. Bush's Council on Bioethics issued a report in 2002 calling for a ban on cloning for reproductive purposes and a four-year moratorium on therapeutic cloning. Some on the panel remain opposed to therapeutic cloning because they worry that it is a stepping-stone to reproductive cloning. Others on the council struggled between their "sympathy for the sick" and their "piety for human life." The council has come under heavy criticism. Bioethicist Arthur Caplan believes that the council "will do nothing to jostle any of the president's already espoused positions condemning stem cell research, cloning, and the creation of human embryos for research." (24) But because this technique is not creating a new human being, cancer researcher Bert Vogelstein, President of the National Academy of Sciences Bruce Alberts, and President of The Institute of Medicine Kenneth Shine argue that the term therapeutic cloning is "conceptually inaccurate and misleading, and should be abandoned." As an alternative, they propose the term "nuclear transplantation," which they believe "captures the concepts of the cell nucleus and its genetic material being moved from one cell to another, as well as the nuance of 'transplantation,' an objective of regenerative medicine." (25)

This is an issue that will take time to resolve. But that people are talking about and debating science like cloning, nuclear transplantation, and DNA fingerprinting and its impact on society is a good thing. This will hopefully lead to a public that are more engaged and interested in the ways in which new technologies will affect their lives. It should also help the public weigh the risks and benefits of the application of genomic technologies. It will still be years if not decades until genomics becomes a routine part of our lives. After all, the genomic revolution itself is just getting off the ground. Eric Lander, one of the leaders of the Human Genome Project and head of the Whitehead Institute for Biomedical Research at the Massachusetts Institute of Technology, sees the completion of the sequence of the human genome as a first step. In April 2003, celebrating the completion of NHGRI's final draft of the human genome sequence, Lander acknowledged that "starting today, the real serious analysis of things can begin." (26)

The completed genome sequence now yields to the continuing task of making sense of our biology in a way that can help prevent and cure disease, help us better understand the evolutionary heritage of life on Earth, and find ways to safely enhance agriculture. In the preceding eight chapters we have introduced the fundamentals of genomics. From the discovery of the structure of DNA to the sequencing of the human genome to the integration of biotechnology into health care, genomics has and will continue to have an important impact on all of our lives. It would be a terrible loss if the side effects of genomic technology some day outshine its benefits. Now, at the outset of the genomic revolution, it is the responsibility of all of us to make sure that the best of this new science is made part of our lives.

An Experiment: Seeing Your Own DNA

Did you know that DNA is visible if you have enough of it in one place? It is actually really easy to isolate DNA from your own cells; you only need some common kitchen items and utensils to see your own DNA. The first step in seeing your own DNA is finding a bunch of your cells.

Step 1: Make a glass of salt water and swish it around in your mouth for 30 seconds. This should be enough to dislodge some of your cells from the inside of your cheeks (as well as thousands of bacterial cells). The amount of salt in the water should be the same as if you were trying to help a sore throat by gargling.

Step 2: Spit the swished salt water from your mouth into a thin glass. A very thin or small glass works fine. If you look closely at the salt water in the glass you will see it is very cloudy. This is because there are literally hundreds of thousands of cells (yours and those of different bacteria) in the water. Because the DNA is on the inside of the cells we need a way to break open the cells. A very thin membrane called a cell membrane surrounds our cells. These membranes are made up of lipids, which are fat molecules. Think hard now, how do you disperse fatty substances. You or someone in your family does it every evening when you wash the dishes. SOAP!!

Step 3: Add several drops of liquid soap to the salty water with your cells in it (dish soap works just fine). Gently swirl the glass. You might notice the water in the glass getting clearer. This is because the soap is actually breaking the cell membranes apart, literally blowing the cells open and releasing into the salt water all the material from inside the cells. The water clears up because the blown-apart cells are now dissolving in the salty water. Now we have a salty mixture of cellular materials that includes DNA from your chromosomes and the chromosomes of thousands of bacteria. There are also a lot of proteins in the mixture, too.

169

Step 4: Slowly pour alcohol into the salt water so that a layer of alcohol forms on top of the salt water. Rubbing alcohol works best for this, but other alcohols like vodka, tequila or rum will work, too. We are taking advantage of the fact that nucleic acids "hate" alcohol. When DNA is in the presence of alcohol it rushes away from the ethanol and rapidly comes out of solution.

Step 5: Let the solution sit for a few minutes. If there is enough DNA you will shortly see a bunch of stringy stuff rising away from the salty layer into the alcohol layer. The stringy DNA you see in the glass in front of you is really a collection of the DNA from hundreds of thousands of cells from you and from the bacteria that live in your mouth.

GLOSSARY

Allele—One possible form of a specific gene found at a specific locus on a chromosome.

Amino Acid—The building block of proteins. There are 20 amino acids that make up proteins.

Belmont Report—Issued in 1979 by the National Commission for the Protection of Human Subjects, the Belmont Report established basic regulations for protecting human research subjects based on three governing principles—respect for persons, beneficence, and justice.

Bioethics—The study of the moral and ethical implications of new medical and scientific technologies.

Bioinformatics—An important component of the genomic revolution that utilizes computer science to study and compare genomes. Bioinformaticians develop computer programs to read gene sequences, locate genes, and compare sequences of the same gene in different species.

Biolistics—The crudest but most common method of gene transfer in plants, which utilizes a gene gun to shoot plant tissue with tiny gold or tungsten particles coated with the desired transfer DNA, or T-DNA.

Chargaff's Rule—In all organisms, the ratio of adenine to thymine and the ratio of guanine to cytosine are always 1.

Chromosome—Chromosomes are like genetic scaffolding—they hold in place the long, linearly arranged sequences of deoxyribonucleic acids. Humans have 22 pairs of chromosomes and two sex-determining chromosomes for a total of 46.

Clone—An identical genetic copy of an individual cell or an entire organism.

Codon—A group of three nucleotides (a triplet) that codes for a single amino acid.

Comparative Genomics—A field that identifies genetic differences between species to investigate our planet's evolutionary history and to identify genes and their functions.

171

Welcome to the Genome, by Rob DeSalle and Michael Yudell.
ISBN: 0-471-45331-5 Copyright © 2005 Rob DeSalle and Michael Yudell.

Contig—A continuous stretch of genomic sequences connected from small shotgun sequences or longer BAC or YAC clone sequences.

DNA—Deoxyribonucleic acid, or DNA, is the genetic material that comprises genes. In our cells, DNA is normally shaped like a double helix.

DNA Polymerase—An enzyme that synthesizes DNA. In other words, DNA polymerase is the mechanism by which DNA clones or copies itself.

Dominant Trait—A genetic variation that always appears, even in the presence of what is known as a recessive trait.

Double Helix—DNA, made up of two interconnected chains of nucleic acids, is normally shaped like a spiral staircase or double helix.

ELSI (the Ethical, Legal, and Social Implications Program of the Human Genome Project)—Formally established in 1990, ELSI has dispensed over $100 million in research funds to academics and policy makers to study the ethical, legal, and social implications of genomic research. It is the largest bioethics program ever established.

Enzymes—A class of proteins that usually accelerate chemical reactions in an organism.

Eugenics—The belief that selective breeding can improve the "genetic quality" of humans.

Evolution—Classically defined as a change in allele frequency with time.

Exon—The coding regions of DNA within a gene.

Gene-Environment Interaction—The vast majority of all human disease is the result of the interaction between genes, organism, and environment.

Genes—Genes are made up of DNA and are the basic units of inheritance in all living organisms.

Gene Therapy—The practice of fixing genetic malfunctions by either delivering working genes into an organism's cells or repairing or replacing malfunctioning genes. There are two categories of gene therapy: Somatic cell gene therapy repairs or replaces a malfunctioning or missing gene in somatic or nonreproductive cells, and germline therapy does the same in germline or reproductive cells.

Genetically Modified Organism (or "GMO")—GMOs have a working copy of a gene inserted into their genome from another variety of the same species or from a foreign species. This can be done through a variety of methods including biolistics, electoporation, and bacterial gene transport.

Genetic Counselor—Genetic counselors work closely with people being genetically tested to ensure that before to testing they are informed of all the ramifications of a genetic test, that they are given the appropriate genetic test once they make an informed decision to proceed, and that the testing information is correctly interpreted.

Genetic Discrimination—A scenario in which the results of genetic testing could be used either to deny health insurance based on a preexisting condition or to make employment decisions adverse to the worker. Many states prohibit genetic discrimination, and the federal government is currently considering anti-genetic discrimination legislation.

Genetic Enhancement—The engineering of genes to improve upon selected human traits, making people somehow better than normal.

Genetic Test—Tests on a person's DNA that can unveil associations between genes and the diseases they either cause or contribute to. As of October 2003, there were 993 genetic tests that identify genetic causes of disease, of which 645 are available for clinical use. Genetic tests can be conducted by using microarrays, some form of DNA sequencing, or tests of blood samples for abnormal or missing gene products.

Genome—The entire set of an organism's genetic material.

Genomics—Genomics is all about looking at hundreds, if not thousands, of genetic interactions simultaneously in order to understand the root causes of disease and to better understand how an organism works.

Genotype—An organism's genetic makeup.

Germ Cells—Reproductive cells that contain a single copy of the human genome. In humans germ cells contain 23 nonpaired chromosomes.

Haplotype—Set of genetic material that is inherited in blocks of between 5000 and 200,000 bases. These haptotype blocks may be a key to understanding disease-related genomic information.

Heredity—The ways in which traits are passed between generations via genes.

HOX Genes—A group of genes involved in the development of basic animal body structure.

Human Genome Project—Officially begun in 1990, the Human Genome Project is an international effort to map, sequence, and understand the human genetic code. The project officially completed the human genome sequence in April 2003.

Hybridization—A chemical reaction in which a strand of DNA or RNA reacts with another strand of DNA or RNA to produce a double helix. The better the complementarity (A matching with T or U, C matching with G), the better the reaction.

Informed Consent—United States federal regulations require that participants in scientific and medical research give their informed consent by affirmatively acknowledging that they understand the nature of the research and are told the risks and benefits of participation. Unlike most basic research, genetic research can involve risks that continue past the completion of a study.

Intron—DNA region within a gene that does not code for a protein. Some scientists hypothesize that these noncoding regions may play a role in regulating gene function.

Junk DNA—A term originally used to describe DNA that did not code for a protein or structural RNA. The term may turn out to be a misnomer. Some scientists hypothesize that these noncoding regions of DNA may play a role in regulating gene function.

Locus or Loci—A position of a gene on a chromosome that can contain a gene or its alleles.

Messenger RNA—The intermediary between genes and proteins, messenger RNA, or mRNA, takes genetic instructions from DNA to ribosomes, where genetic information is decoded and a protein is produced.

Microarray—A genomic technology that allows researchers to look at the activity of many individual genes in a cell and also to look at the interactions of as many as thousands of genes at a time within a cell. Microarrays have many applications, from locating genes that are potential targets for pharmaceutical development to pinpointing the genetic roots of diseases to studying genetic responses to environmental toxins. A micorarray is generally either a simple glass microscope slide or a piece of nylon affixed to plastic.

Mitochondrion—A cellular structure responsible for the production of energy within a cell. A small portion of the human genome is found in the mitochondrion. The human mitochondrial genome contains 16,569 bases and 37 genes.

Multigenic Trait—A trait that is influenced by more than one gene.

Mutation—A change in an organism's DNA sequence that can be a random event or caused by an external factor such as an environmental toxin or radioactivity.

Nucleic Acid—Composed of a nucleotide base, sugar, and phosphoric acid. Makes up the most basic building block of life—DNA.

Nucleotides—There are four different nucleotides that make up DNA—adenine, thymine, guanine, and cytosine. These four nucleotides are commonly referred to as A, T, G, and C.

Nuremberg Code—Written by the judges at the Nazi doctors' trial in Nuremberg, the Code is the foundation for modern medical ethics, laying out 10 directives for protecting human research subjects. At the core of the Code is the principle that "voluntary consent of the human subject is absolutely essential" in a research protocol.

Pharmacogenomics—The application of genomic technology to the creation of pharmaceuticals.

Phenotype—The observable characteristics of an organism are its phenotype.

Polygenic—When more than one gene is involved in the expression of a trait it is known as polygenic.

Polygeny—The idea that the human races were created in a hierarchy of separate species.

Polymerase Chain Reaction (PCR)—A laboratory procedure devised by Kerry Mullis and his colleagues that clones (copies) a particular fragment of DNA.

Preimplantation Genetic Diagnosis—A method that allows doctors to genetically test embryos in vitro and select a desired embryo based on a specific or multiple traits. The method has been used in limited cases to select an embryo that is a donor match for a sick sibling.

Protein—Proteins carry out the varied instructions inscribed in your DNA.

Proteomics—The science of decoding the human proteome, the entire collection of the approximately 500,000 human proteins.

Recessive Trait—A genetic variation that cannot manifest itself when a dominant trait is present.

Ribosomes—Cellular structures where protein synthesis occurs.

RNA—Ribonucleic acids, or RNA, are one of two types of nucleic acids in cells. Messenger RNA carries genetic instructions between DNA and proteins. RNA is made up of four bases: adenine, cytosine, guanine, and uracil.

Sequencing—The linear decoding of a stretch of DNA or an entire organism's genome.

Single Nucleotide Polymorphism (or "SNP")—In every 1000-base (GATC) stretch in the human genome there is approximately one difference between you and someone who is not your close relative. These differences are known as SNPs.

Somatic Cells—The cells that make up the human body. Unlike germ cells, they are not capable of being fertilized for human reproduction. Somatic cells contain two complete copies of the human genome.

Stem Cell—A type of cell that is active throughout the life of an organism in making new cells and other tissue. Stem cells in bone marrow, for example, are constantly making new blood cells. Some stem cells, including embryonic stem cells, can differentiate into specialized cells found in all body tissue.

Toxicogenomics—A new field that is examining the human genome to understand how genes are affected by environmental toxins.

Trait—The physical manifestation of the interaction of genes and environment. Your height and weight are traits.

Transgenic—Refers to an organism that has had genetic material inserted into its genome.

Transcription—The copying of the DNA code into mRNA. mRNA molecules take genetic information to ribosomes.

Translation—The process by which mRNA is translated into proteins at ribosomes.

Tree of Life—The international effort to map evolutionary relationships of all named species (1.7 million) on our planet.

ENDNOTES

Introduction

1. Nicolas Wade. 1999. "Long Held Beliefs Are Challenged by New Human Genome Analysis," *New York Times* (August 3, 1999) p.A20; Natalie Angier. 2001. "Genome Shows Evolution Has an Eye for Hyperbole,"*New York Times* (February 13, 2001) p.F1; and Geoff Dyer, David Firn, and Victoria Griffith. 2003. "Double Helix Is Starting To Make Its Mark In Medicine," *Financial Times (London)* (July 4, 2003) p.20.

2. Euripides, *Electra*, as quoted in Conway Zirkle. 1951. "The Knowledge of Heredity Before 1900," *Genetics in the Twentieth-Century: Essays on the Progress of Genetics During Its First 50 Years*. L.C. Dunn editor, New York: The MacMillan Company, p.42.

3. Hans Stubbe. 1972. *History of Genetics: From Prehistoric Times to the Rediscovery of Mendel's Laws*. Cambridge, MA: The MIT Press, pp.51–52.

4. Stubbe, 1972, pp.1–6.

5. Stubbe, 1972, p.33.

6. Ernst Mayr. 1982. *The Growth of Biological Thought: Diversity, Evolution, and Inheritance*. Cambridge, MA: The Belknap Press of the Harvard University Press, pp.636–637; and Ernst Mayr. 1988. *Toward a New Philosophy of Biology: Observations of an Evolutionist*. Cambridge, MA: The Belknap Press of Harvard University Press.

7. Richard Saltus. 2000. "Decoding of Genome Declared," *Boston Globe* (June 27, 2000) p.A1.

8. Nicolas Wade. 2000. "Reading the Book of Life," *New York Times* (June 27, 2000) p.A1.

9. Sabin Russell. 2002. "Researchers Optimistic on Cancer Drugs," *San Francisco Chronicle* (February 19, 2002) p.A3.

10. Kenneth Olden and Samuel Wilson. 2000. "Environmental Health and Genomics: Visions and Implications," *Nature Reviews Genetics* 1:149–153.

Chapter 1

1. Wojciech Makalowski. 2003. "Not Junk After All," *Science* 300:1246–1247.

2. S. Anderson et al. 1981. "Sequence and Organization of the Human Mitochondrial Genome," *Nature* 290:457–465; http://www.mitomap.org/mitomap/mitoseq.html.

3. Kip A. West et al. 2003. "Rapid Akt Activation by Nicotine and a Tobacco Carcinogen Modulates the Phenotype of Normal Human Airway Epithelial Cells," *Journal of Clinical Investigation* 111:81–90; Kristine Novak. 2003. "Double Whammy," *Nature Reviews Cancer* 3:83.

4. Peter J. Bowler. 1989. *The Mendelian Revolution: The Emergence of Hereditarian Concepts in Modern Science and Society.* Baltimore: The Johns Hopkins University Press.

5. Stephen Jay Gould. 1977. *Ontogeny and Phylogeny.* Cambridge, MA: The Belknap Press of the Harvard University Press, pp.19–20.

6. Vítezslav Orel. 1984. *Mendel.* New York: Oxford University Press, pp.19–23.

7. Robin Marantz Henig. 2000. *The Monk in the Garden: The Lost and Found Genius of Gregor Mendel, the Father of Genetics.* New York: A Mariner Book, pp.21–22.

8. Orel, 1984, pp.28–33.

9. Mayr, 1982, p.725.

10. Henig, 2000, pp.69–93.

11. Henig, 2000, p.86.

12. Henig, 2000, pp.85–86.

13. D. Peter Snustad and Michael J. Simmons. 2003. *Principles of Genetics.* New York: John Wiley & Sons, p.55.

14. Henig, 2000, p.140.

15. Orel, 1984, p.51.

16. Henig, 2000, p.140.

17. Henig, 2000, p.79.

18. Orel, 1984, p.92.

19. Ulf Lagerkvist. 1998. *DNA Pioneers and Their Legacy.* New Haven: Yale University Press, pp.101–102.

20. Orel, 1984, p.93.

21. Bowler, 1989, p.3; EJ. Brown. 1995. *Charles Darwin: A Biography.* New York: Knopy Adrian Desmond. 1997. *Huxley: From Devil's Disciple to Evolution's High Priest.* Reading, MA: Addison-Wesley.

22. Mayr, 2002, p.727.

23. Bowler, 1989, pp.108–116.

24. Garland E. Allen. 1978. *Thomas Hunt Morgan: The Man and His Science.* Princeton: Princeton University Press, p.165.

25. L.C. Dunn. 1965. *A Short History of Genetics: The Development of Some of the Main Lines of Thought: 1864–1939.* New York: McGraw-Hill Book Company, pp.xvii, xxi.

26. Allen, 1978, p.149.

27. Allen, 1978, pp.150–153.

28. Donald Pierce, et al. 1996. "Studies of the Mortality of Atomic Bomb Survivors, Report 12, Part 1. Cancer: 1950–1990," *Radiation Research* 146:1–27; A.M. Kellerer. 2000. "Risk

Estimates for Radiation-Induced Cancer–The Epidemiological Evidence," *Radiation and Environmental Biophysics* 39:17–24.

29. Neil A. Campbell. 1999. *Biology*. New York: Addison Wesley Longman, Inc., p.359.

30. Allen, 1978, p.172.

31. Allen, 1978, pp.173–179.

32. Allen, 1978, pp.175–176.

33. Allen, 1978, p.164

34. Allen, 1978, pp.173–179.

35. Daniel J. Kevles. 1995. *In the Name of Eugenics: Genetics and the Uses of Human Heredity*. Cambridge, MA: Harvard University Press, pp.3–19.

36. Francis Galton. 1883. *Inquiries into Human Faculty and its Development*. London: J.M. Dent, p.17.

37. Joe William Trotter, Jr. 1991. *The Great Migration in Historical Perspective: New Dimensions of Race, Class, and Gender*. Bloomington: Indiana University Press; Nicolas Lemann. 1992. *The Promised Land: The Great Black Migration and How it Changed America*. New York: Vintage Books; Matthew Jacobson. 1998. *Whiteness of a Different Color: European Immigrants and the Alchemy of Race*. Cambridge, MA: Harvard University Press.

38. Charles B. Davenport. 1923. *Eugenics, Genetics and the Family: Volume 1, Scientific Papers of the Second International Congress of Eugenics*, Baltimore: Williams & Wilkins Company, p.4.

39. Allen, 1978, p.232.

40. William H. Tucker. 1994. *The Science and Politics of Racial Research*. Urbana: University of Illinois Press, p.95; Kevles, 1995, pp.102–103.

41. Celeste Michelle Condit. 1999. *The Meanings of the Gene: Public Debates About Human Heredity*. Madison: University of Wisconsin Press, p.27.

42. Elof Axel Carlson. 2001. *The Unfit: A History of a Bad Idea*. Cold Spring Harbor, NY: Cold Spring Harbor Laboratory Press, p.12.

43. Paul Popenoe and Rosewill Hill Johnson. 1933. *Applied Eugenics*. New York: MacMillam, p.141.

44. Kevles, 1995, p.46.

45. William Provine. 1986. "Genetics and Race," *American Zoologist* 26: 857–887.

46. As quoted in Provine, 1986, p.868.

47. Madison Grant. 1916. *The Passing of the Great Race: The Racial Basis of European History*. New York: Charles Scribner and Sons, p.45.

48. William H. Tucker. 2002. *The Funding of Scientific Racism: Wickliffe Draper and the Pioneer Fund*. Chicago: University of Illinois Press, p.27.

49. Stefan Kühl. 1994. *The Nazi Connection: Eugenics, American Racism, and German National Socialism*. New York: Oxford University Press, p.39.

50. Edwin Black. 2003. *War Against the Weak: Eugenics and America's Campaign to Create a Master Race*. New York: Four Walls Eight Windows, pp.294–295, 313–314.

51. Black, 2003, p.312.

52. Black, 2003, p.314.

53. Kühl, 1994, pp.48–49.

54. David Micklos and Elof Carlson. 2000. "Engineering American Society: The Lesson of Eugenics," *Nature Reviews Genetics* 1:153–158.

55. Theodosius Dobzhansky. 1982. *Genetics and the Origin of Species*. New York: Columbia University Press, pp. xvii, 118–121.

56. Mayr, 1982, pp.567–570.

57. Michel Morange. 1998. *A History of Molecular Biology*. Cambridge, MA: Harvard University Press, p.34.

58. Oswald T. Avery, Colin M. MacLeod, and Maclyn McCarty. 1944. "Studies on the Chemical Nature of the Substance Inducing Transformation of Pneumococcal Types," *Journal of Experimental Medicine* 79:137–158.

59. Allen, 1978, p.208; Mayr, 1982, p.818

60. Erwin Chargaff. 1971. "Preface to a Grammar of Biology: A Hundred Years of Nucleic Acid Research," *Science* 172:637–642; Mayr, 1982, p.819.

61. Robert Olby. 1974. *The Path to the Double Helix*. Seattle: University of Washington Press, pp.211–221.

62. James Watson. 1968. *The Double Helix: A Personal Account of the Discovery of the Structure of the Double Helix*. New York: Atheneum, pp.23–24.

63. Olby, 1974, pp.381–383.

64. Brenda Maddox. 2002. *Rosalind Franklin: The Dark Lady of DNA*. New York: HarperCollins Publishers, p.191.

65. J.D. Bernal, 1958. "Obituary Notice of Rosalind Franklin," *Nature* 182:154, as quoted in Maddox, 2002. p. xviii.

66. Watson, 1968, pp.167–168.

67. James Watson and Francis Crick. 1953. "Molecular Structure of Nucleic Acids: A Structure for Deoxyribonucleic Acid," *Nature* 171:737–738; Maddox, 2002, pp.196–197.

68. Maddox, 2002, pp.311–312.

69. Watson, 1968, pp. 17, 165–166.

70. Maddox, 2002, p.202.

71. Evelyn Fox Keller. 2000. *The Century of the Gene*. Cambridge, MA: Harvard University Press, p.2.

Chapter 2

1. Erwin Schrödinger. 1992. *What is Life: The Physical Aspect of the Living Cell*. New York: Cambridge University Press, p.5.

2. Schrödinger, 1992, p.6.

3. Schrödinger, 1992, p.3.

4. Schrödinger, 1992, p.5

5. Horace Freeland Judson. 1996. *The Eighth Day of Creation: Makers of the Revolution in Biology*. Cold Spring Harbor, NY: Cold Spring Harbor Laboratory Press, pp.29, 87–88; James D. Watson and Andrew Berry. 2003. *DNA: The Secret of life*. New York: Alfred A. Knopf, pp.35–36.

6. Stephen Jay Gould. 1995. "'What is Life?' As a Problem in History," *What is Life? The Next Fifty Years: Speculations on the Future of Biology*. Michael P. Murphy and Luke A.J. O'Neill, eds. Cambridge: Cambridge University Press, p.26.

7. Frederick Sanger. 1945. "The Free Amino Groups of Insulin," *Biochemical Journal* 39:507.

8. _____. 2003. "What You Need to Know About. . . Phenylketonuria," *Nursing Times* 99:26.

9. Frederick Sanger. 1959. "Chemistry of Insulin," *Science* 129: 1340–1344; Frederick Sanger. 1988. "Sequences, Sequences, Sequences," *Annual Review of Biochemistry* 57:1–28.

10. Sanger, 1959, pp.1340–1344.

11. Judson, 1996, p. 88.

12. Sanger, 1959, pp.1340–1344.

13. Judson, 1996, pp.256–266

14. Judson, 1996, p.470.

15. Michael Morange. 1998. *A History of Molecular Biology*. Cambridge, MA: Harvard University Press, p.135; Marhsall W. Nirenberg and Johann H. Matthaei. 1961. "The Dependence of Cell-Free Protein Synthesis in *E. coli* upon Naturally Occurring or Synthetic Polyribonucleotides," *Proceedings of the National Academy of Sciences USA* 47:1588–1602.

16. Judson, 1996, p.298.

17. Horace Freeland Judson. 1992. "A History of the Science and Technology Behind Gene Mapping and Sequencing," in *The Code of Codes: Scientific and Social Issues in the Human Genome Project*. Daniel J. Kevles and Leroy Hood, eds. Cambridge, MA: Harvard University Press, p.59.

18. Sydney Brenner, François Jacob, and Matthew Messelson. 1961. "An Unstable Intermediary Carrying Information From Genes to Ribosomes for Protein Synthesis," *Nature* 190:576–581; Françoise Jacob and Jacque Monod. 1961. "Genetic Regulatory Mechanisms in the Biosynthesis of Proteins," *Journal of Molecular Biology* 3:318–356.

19. Morange, 1998, p.22.

20. George W. Beadle. 1977. "Genes and Chemical Reactions in Neurospora," in *Nobel Lectures in Molecular Biology, 1933–1975*. New York: Elsevier, pp.51–63; Tonse N.K. Raju. 1999. "The Nobel Chronicles," *The Lancet* 353:2082.

21. Arthur Kornberg. 1989. *For the Love of Enzymes: The Odyssey of a Biochemist*. Cambridge, MA: Harvard University Press, p.121.

22. Kornberg, 1989, pp.147–154.

23. Morange, 1998, pp.236–237.

24. Kornberg, 1989, pp.217–220, 240–268.

25. Edward Southern. 1975. "Detection of Specific Sequences Among DNA Fragments Separated By Gel Electrophoresis," *Journal of Molecular Biology* 98:503–517.

26. David C. Darling and Paul M. Brickell. 1994. *Nucleic Acid Blotting: The Basics*. New York: Oxford University Press.

27. Stanley Cohen et al. 1973. "Construction of Biologically Functional Bacterial Plasmids *in vitro*," *Proceedings of the National Academy of Sciences USA* 70:3240–3244.

28. Eric Green. 2002. "Sequencing the Human Genome: Elucidating Our Genetic Blueprint," in *The Genomic Revolution: Unveiling the Unity of Life*. Michael Yudell and Rob DeSalle eds. Washington, D.C.: Joseph Henry Press, p.39.

29. Robert Cook-Deegan. 1994. *The Gene Wars: Science, Politics, and the Human Genome*. New York: W.W. Norton and Company, p.62.

30. Kevin Davies. 2001. *Cracking the Genome: Inside the Race to Unlock Human DNA*. New York: The Free Press, p.37.

31. Frederick Sanger et al. 1977. "DNA Sequencing with Chain-Terminating Inhibitors," *Proceedings of the National Academy of Sciences USA* 74:5463–5467.

32. Green, 2002, p.40; Sanger, 1988, pp.1–28.

33. Cook-Deegan, 1994, p.62.

34. Morange, 1998, p.205.

35. International Human Genome Sequencing Consortium. 2001. "Initial Sequencing and Analysis of the Human Genome," *Nature* 409: 860–921; J. Craig Venter et al. 2001. "The Sequence of the Human Genome," *Science* 291:1304–1351.

36. Kary Mullis. 1990. "The Unusual Origin of the Polymerase Chain Reaction," *Scientific American* (April 1990) pp.56–65.

37. Kary Mullis et al. 1986. "Specific Enzymatic Amplification of DNA in Vitro: The Polymerase Chain Reaction," *Cold Spring Harbor Symposium in Quantitative Biology* 51:263–273.

38. Paul Rabinow. 1996. *Making PCR: A Story of Biotechnology*. Chicago: University of Chicago Press, pp.128–129.

Chapter 3

1. For two different histories of the genome project take a look at Kevin Davies. 2001. *Cracking the Genome: Inside the Race to Unlock Human DNA*. New York: The Free Press; and John Sulston and Georgina Ferry. 2002. *The Common Thread: A Story of Science, Politics, Ethics and the Human Genome*. Washington, DC: Joseph Henry Press.

2. D. Peter Snustad and Michael J. Simmons. 2003. *Principles of Genetics: 3rd Edition*. New York: John Wiley & Sons, Inc., p.20.

3. Jeremy Nathans, Darcy Thomas, and David S. Hogness. 1986. "Molecular Genetics of Human Color Vision: The Genes Encoding Blue, Green and Red Pigments," *Science* 232:193–202.

4. Jeremy Nathans, Thomas P. Piantanida et al. 1986. "Molecular Genetic of Inherited Variation in Human Color Vision," *Science* 232:203–210.

5. Takaaki Hayashi et al. 1999. "Position of a 'Green-Red' Hybrid Gene in the Visual Pigment Array Determines Colour-Vision Phenotype," *Nature Genetics* 22:90–93.

6. Jeremy Nathans, Thomas P. Piantanida et al., 1986, pp.203–210

7. Hai Yan et al. 2001. "Small Changes in Expression Affect Predisposition to Tumorigenesis," *Nature Genetics* 30:25–26.; Kenneth W. Kinzler and Bert Vogelstein. 1996. "Lesson From Hereditary Colon Cancer," *Cell* 87:159–170.

8. National Research Council, Committee on Mapping and Sequencing the Human Genome. 1998. *Mapping and Sequencing the Human Genome.* Washington, DC: National Academy Press, p.65.

9. Robert Cook-Deegan. 1994. *The Gene Wars: Science, Politics, and the Human Genome.* New York: W.W. Norton and Company, p.79.

10. Cook-Deegan, 1994, pp.84, 88–89.

11. James D. Watson. 1990. "The Human Genome Project: Past, Present, and Future," *Science* 248:44–49.

12. Cook-Deegan, 1994, p.171.

13. Dorothy Nelkin. 1992. "The Social Power of Genetic Information," *The Code of Codes: Scientific and Social Issues in the Human Genome Project.* Daniel J. Kevles and Leroy Hood, eds. Cambridge, MA: Harvard University Press, pp.177–190.

14. National Research Council, 1988, pp.5–6.

15. HUGO Mission Statement: *http://www.hugo-international.org/hugo/HUGO-mission-statement.htm*; Victor A. McKusick. 1989. "The Human Genome Organization: History, Purposes and Membership," *Genomics* 5:385–387.

16. National Research Council, 1988, p.9.

17. Cook-Deegan, 1994, pp.64–72.

18. Meredith W. Salisbury, "Four Color Face Off," *Genome Technology* (June 2002) p.70–75.

19. Davies, 2001, pp.145–146.

20. Francis S. Collins et al. 1998. "New Goals for the U.S. Human Genome Project: 1998–2003," *Science* 282:682–689.

21. Nicolas Wade. 1998. "Scientist's Plan: Map All DNA Within 3 Years," *New York Times* (May 10, 1998) p.1.

22. Meredith Wadman. 1998. "Company Aims to Beat NIH Human Genome Efforts," *Nature* 393:101; Davies, 2001, p.148.

23. Eliot Marshall. 2001. "Bermuda Rules: Community Spirit, With Teeth," *Science* 291:1192; Davies, 2001, p.87.

24. Wade, 1998, p.1.

25. Richard Preston. 2000. "The Genome Warrior" *The New Yorker* (June 12, 2000) pp.66–83.

26. Leslie Roberts, 1991. "Gambling on a Shortcut to Genome Sequencing," *Science* 252: 1618–1619.

27. Leslie Roberts. 1991. "Genome Patent Fight Erupts," *Science* 254:184–186.

28. Roberts, 1991, p.184.

29. Preston, 2000, p.71.

30. An Introduction to *C. elegans*: *http://www.biotech.missouri.edu/Dauer-World/Wormintro.html*

31. J. Craig Venter. 2002. "Whole-Genome Shotgun Sequencing," *The Genomic Revolution: Unveiling the Unity of Life*. Michael Yudell and Rob DeSalle, eds. Washington, DC: Joseph Henry Press, pp.49–50; Preston, 2000, p.72; Hamilton Smith et al. 1995. "Frequency and Distribution of DNA Uptake Signal Sequences in the Haemophilus Influenzae Rd Genome," *Science* 269:538–540.

32. J. Craig Venter et al. 1998. "Shotgun Sequencing of the Human Genome," *Science* 280:1540–1542; Wadman, 1998, p.101.

33. J. Craig Venter et al. 2001. "The Sequence of the Human Genome," *Science* 291:1304–1351; Greg Gibson and Spencer V. Muse. 2002. *A Primer of Genome Science*. Sunderland, MA: Sinauer Associates, Inc. Publishers, pp.78–91.

34. International Human Genome Sequencing Consortium. 2001. "Initial Sequencing and Analysis of the Human Genome," *Nature* 409: 860–921; Robert H. Waterson, Eric S. Lander, and John E. Sultson. 2002. "On Sequencing the Human Genome," *Proceedings of the National Academy of Sciences USA* 99:3712–3716; Jennifer Couzin. 2002. "Taking Aim at Celera's Shotgun," *Science* 295:1817.

35. Press Release, The Wellcome Trust, "The Finished Human Genome—Wellcome to the Genomic Age," April 14, 2003. *http://www.sanger.ac.uk/Info/Press/2003/030414.shtml*

36. Davies, 2001, p.164; National Human Genome Research Institute website: *http://www.genome.gov/10001674#sequencing*.

37. Nicolas Wade. 1998. "New Company Joins Race to Sequence the Human Genome," *New York Times* (August 18, 1998) p.F6.

38. William Allen. 2000. "Competition Has Accelerated Race To Sequence The Genome," *St. Louis Post-Dispatch* (June 4, 2000) p.A11.

39. Preston, 2000, p.66.

40. National Human Genome Research Institute: *http://www.genome.gov/10000779*

41. Davies, 2001. p.148.

42. Andy Coghlan and Nell Boyce. 2000. "The First Draft of the Human Genome Signals a New Era for Humanity," *New Scientist* (July 1, 2000) p.44; Frederic Golden and Michael D. Lemonick. 2000. "The Race is Over," *Time* (July 3, 2000).

43. Nicolas Wade. 2000. "Reading the Book of Life," *New York Times* (June 27, 2000) p.A1.

44. Adrienne Burke et al. 2002. "Venter Uncut," *Genome Technology* (December 2002) pp.38–41; Press Release. 2002. "J. Craig Venter, Ph.D., Announces Formation of Three Not-for-Profit Organizations Will Focus on Ethical and Social Issues Surrounding Genomics and Developing New Biological Energy Sources," *http://www.tcag.org/news.html*.

Chapter 4

1. National Science Board. 2002. *Science and Engineering Indicators–2002*. Arlington, VA: National Science Foundation.

2. A November 2002 Harris Interactive Poll shows that 77% of Americans generally trust doctors to tell the truth. The number is slightly lower for scientists—68%. Humphrey Taylor. 2002. "Trust in Priests and Clergy Falls 26 Points in Twelve Months," *The Harris Poll* #63(November 27, 2002) or *http://www.harrisinteractive.com/harris_poll/index.asp?PID=342*; National Science Board. 2002. *Science and Engineering Indicators–2002*. Arlington, VA: National Science Foundation.

3. David J. Rothman and Sheila M. Rothman. 2002. "Redesigning the Self: The Promise and Perils of Genetic Enhancement," *The Genomic Revolution: Unveiling the Unity of Life*. Washington, DC: Joseph Henry Press, p.155.

4. Jonathan D. Moreno. 2000. *Undue Risk: Secret State Experiments on Humans*. New York: W. H. Freeman and Company, pp.54, 58.

5. Robert Proctor. 1988. *Racial Hygiene: Medicine Under the Nazis*. Cambridge, MA: Harvard University Press; Robert Jay Lifton. 1986. *The Nazi Doctors: Medical Killing and the Psychology of Genocide*. New York: Basic Books.

6. Moreno, 2000, p.61.

7. Arthur Caplan. 1992. "How Did Medicine Go So Wrong?" *When Medicine Went Mad: Bioethics and the Holocaust*. Arthur Caplan, ed. Totowa, NJ: Humana Press, pp.53–92.

8. Nancy L. Segal. 1992. "Twin Research at Auschwitz-Birkenau: Implications for the Use of Nazi Data Today," *When Medicine Went Mad: Bioethics and the Holocaust*. Arthur Caplan, ed. Totowa, NJ: Humana Press, pp.281–299.

9. Segal, 1992, p.284.

10. The entire Nuremberg Code can be found on-line at *http://ohsr.od.nih.gov/nuremberg.php3*. The original text is in the proceedings of the Nazi doctors' trial. *Trials of War Criminals Before the Nuremberg Military Tribunals Under Control Council Law No. 10, vol.2*, Washington, DC: U.S. Government Printing Office, 1949, pp.181–182.

11. Henry K. Beecher. 1966. "Ethics and Clinical Research," *New England Journal of Medicine* 274:1354–1360.

12. David J. Rothman. 1992. *Strangers at the Bedside: A History of How Law and Bioethics Transformed Medical Decision Making*. New York: Basic Books, p.53.

13. Beecher, 1966, pp.1354–1360.

14. James H. Jones. 1993. *Bad Blood: The Tuskegee Syphilis Experiment*. New York: The Free Press.

15. Jones, 1993, p.11.

16. Rothman, 1992, p.251.

17. Albert Jonsen. 1998. *The Birth of Bioethics*. New York: Oxford University Press, p.148.

18. James Childress. 2000. "Nuremberg's Legacy: Some Ethical Reflections," *Perspectives in Biology and Medicine* 43:347–361.

19. The National Commission for the Protection of Human Subjects of Biomedical and Behavioral Research. 1979. "The Belmont Report: Ethical Guildelines for the Prtoection of Human Subjects." DHEW publication no. (OS) 78–0012. Washington, DC.

20. Jeremy Rifkin. 1993. "This Research Overpromises," *USA Today*. December 20, 1993, p.12A.

21. John Leo. 1989. "Genetic Advances, Ethical Risks," *US News and World Report* (September 25, 1989), p.59.

22. Jonsen, 1998, p.15; Childress, 2000, pp.347–361.

23. Arthur Allen. 1997. "Policing the Gene Machine: Can Anyone Control the Human Genome Project?" *Lingua Franca* (March 1997) p.30.

24. Eric T. Juengst. 1991. "The Human Genome Project and Bioethics," *Kennedy Institute of Ethics Journal* 1:71–74.

25. Eric T. Juengst. 1994. "Human Genome Research and the Public Interest: Progress Notes From American Science Policy Experiment," *American Journal of Human Genetics* 54:121–128; Eric M. Meslin et al. 1997. "The Ethical, Legal, and Social Implications Program at the National Human Genome Research Institute," *Kennedy Institute of Ethics Journal* 7:291–298.

26. This figure comes from my E-mail correspondence with Joy Boyer, ELSI Research Staff Program Analyst in Genetics and the Humanities, dated August 21, 2003.

27. Philip Reilly. 1997. "Fear of Genetic Discrimination Drives Legislative Interest," *Human Genome News* 8:3–4.

28. National Conference of State Legislators. 2003. "State Nondiscrimination in Health Insurance Laws." *http://www.ncsl.org/programs/health/genetics/ndishlth.htm*

29. Code of Federal Regulations, Title 45 Public Welfare, Department of Health and Human Services, National Institutes of Health, Office for Protection From Research Risks, Part 46, Protection of Human Subjects. (December 13, 2001): *http://ohrp.osophs.dhhs.gov/humansubjects/guidance/45cfr46.htm*

30. Dave Wendler, Kiran Prasad, and Benjamin Wilfond. 2002. "Does the Current Consent Process Minimize the Risks of Genetics Research," *American Journal of Medical Genetics* 113:258–262.

31. Arthur Caplan. 2001. "Mapping Morality: The Rights and Wrongs of Genomics," *The Genomic Revolution: Unveiling the Unity of Life*. Michael Yudell and Rob DeSalle, eds. New York: Joseph Henry Press, pp.193–197.

32. Henry T. Greely. 2001. "Human Genomics Research: New Challenges for Research Ethics," *Perspectives in Biology and Medicine* 44:221–229.

33. Greely, 2001, pp.221–229.

34. Timothy Caulfield. 2002. "Gene Banks and Blanket Consent," *Nature Reviews Genetics* 3:577.

35. First Genetic Trust website. "Dynamic Informed Consent." *http://www.firstgenetic.net/products_icf.html*

36. Richard R. Sharp and Morris W. Foster. 2000. "Involving Study Populations in the Review of Genetic Research," *Journal of Law, Medicine, and Ethics* 28: 41–51; Greely, 2001.

37. Allen, 1997, pp.29–36.

38. Steven Benowitz, 1996. "Ethical Dilemmas," *The Scientist* 10:1.

39. Stanley Cohen et al. 1973. "Construction of Biologically Functional Bacterial Plasmids In-Vitro," *Proceedings of the National Academy of Sciences USA* 70:3240–3244.

40. Paul Berg et al. 1974. "Potential Biohazards of Recombinant DNA Molecules," *Science* 185:303.

41. Charles Weiner. 2001. "Drawing the Line in Genetic Engineering: Self-Regulation and Public Participation," *Perspectives in Biology and Medicine* 44:208–220.

42. Donald S. Fredrickson. 2001. "The First Twenty-Five Years After Asilomar," *Perspectives in Biology and Medicine* 44:170–182; Nancy M. King. 2002. "RAC Oversight of Gene Transfer: A Model Worth Extention," *Journal of Law, Medicine and Ethics* 30:381–389.

43. Sharon Schmickle. 2002. "Genetic Testing Lawsuit Settled," *Star Tribune (Minneapolis)* (May 9, 2002) p.1D.

44. Rick Weiss. 1995. "Gene Discrimination Barred in the Workplace," *Washington Post* (April 7, 1995) p.A3.

45. Reilly, 1977, pp.3–4; Stepanie Armour. 1999. "Could Your Genes Hold You Back?" *USA Today* (May 5, 1999) p.1B.

46. Justin Gillis. 2000. "Clinton Targets Misuse of Gene Data," *Washington Post* (February 9, 2000) p.A19.

47. Aaron Zitner. 2003. "Senate Blocks Genetic Discrimination," *Los Angeles Times* (October 15, 2003) p.16.

48. Karen H. Rothenberg and Sharon F. Terry. 2002. "Before It's Too Late—Addressing Fear of Genetic Information," *Science* 297: 196–197.

49. Kathy L. Hudson, Karen H. Rothenberg, Lori B. Andrews, Mary Jo Ellis Kahn, and Francis C. Collins. 1995. "Genetic Discrimination and Health Insurance: An Urgent Need for Reform," *Science* 270:391–393.

50. Aaron Zitner. 2003. "Senate Close to Passing Bill Banning Genetic Discrimination," *Los Angeles Times* (May 22, 2003) p.30.

51. Pamela Sankar. 2003. "Genetic Privacy," *Annual Review of Medicine* 54:393–407.

52. Dorothy C. Wertz. 2000. "Genetic Testing in the Workplace: The Lawrence Berkeley Labs Case," *GeneLetter* (April 3, 2000).

53. Lori B. Andrews. 2001. *Future Perfect: Confronting Decisions About Genetics.* New York: Columbia University Press, pp.95–96.

54. Nikunj Somia and Inder M. Verma. 2000. "Gene Therapy: Trials and Tribulations," *Nature Reviews Genetics* 1:91–99.

55. Sheryl Gay Stolberg. 1999. "The Biotech Death of Jesse Gelsinger," *The New York Times Magazine* (November 28, 1999) p.137.

56. Rick Weiss and Deborah Nelson. 2000. "Penn Settles Gene Therapy Suit: University Pays Undisclosed Sum to Family of Teen Who Died," *Washington Post* (November 4, 2000) p.A04.

57. Sheryl Gay Stolberg. 2000. "Biomedicine is Receiving New Scrutiny as Scientists Become Entrepreneurs," *New York Times* (February 20, 2000) p.26.

58. Weiss and Nelson, 2000, p.A04.

59. Jocelyn Kaiser. 2002. "Proposed Rules Aim to Curb Financial Conflicts of Interest," *Science* 295:246–247; Department of Health and Human Services. 2003. "Financial Relationships and Interests in Research Involving Human Subject Protection," *Federal Register* (March 31, 2003) 68:15456–15460.

60. James Meek. 2000. "Poet Attempts the Ultimate in Self-Invention By Patenting Her Own Genes," *The Guardian (London)* (February 29, 2000) p.3; Tom Abate. 2000. "Gene-Patent Opponents Not Licked Yet," *San Francisco Chronicle* (March 20, 2000) p.B1.

61. Rebecca Eisenberg. 2002. "How Can You Patent Genes?" *Who Owns Life?* David Magnus et al., eds. New York: Prometheus Books, pp.117–134.

62. Timothy Caulfield, E. Richard Gold, and Mildred K. Cho. 2000. "Patenting Human Genetic Material: Refocusing the Debate," *Nature Reviews Genetics* 1:227–231.

63. Dorothy Nelkin. 2002. "Patenting Genes and the Public Interest," *American Journal of Bioethics* 2:13–14.

64. Lori B. Andrews. 2002. "Genes and Patent Policy: Rethinking Intellectual Property Rights," *Nature Reviews Genetics* 3:803–807.

65. Philip Reilly. 2001. "Legal Issues in Genomic Medicine," *Nature Medicine* 7:268–271.

66. Paul Gringras and Wai Chen. 2001. "Mechanisms for Differences in Monozygous Twins," *Early Human Development* 64:105–117.

67. Richard Lewontin. 2000. *The Triple Helix: Gene, Organism, and Environment*. Cambridge, MA: Harvard University Press, pp.17–18.

68. Daniel Kelves. 2002. "Eugenics, The Genome, and Human Rights," *The Genomic Revolution: Unveiling the Unity of Life*. Michael Yudell and Rob DeSalle, eds. Washington, DC: Joseph Henry Press, pp.147–148.

69. Arthur Jensen. 1969. "How Much Can We Boost IQ and Scholastic Achievement?" *Harvard Educational Review* 39:449–483.

70. Stephen Jay Gould. 1996. *The Mismeasure of Man*. New York: W.W. Norton & Company.

71. Edward O. Wilson. 1978. *On Human Nature*. Cambridge, MA: Harvard University Press, p.43

72. Jane Alfred. 2000. "Tuning in to Perfect Pitch," *Nature Reviews Genetics* 1:3.

73. Siamak Baharloo et al. 2000. "Familiar Aggregation of Absolute Pitch," *American Journal of Human Genetics* 67:755–758.

74. Mary Jeanne Kreek. 2002. "Gene Diversity in the Endorphin System: SNPs, Chips, and Possible Implications," *The Genomic Revolution: Unveiling the Unity of Life*. Michael Yudell and Rob DeSalle, eds. Washington, DC.: Joseph Henry Press, p.101.

75. Margret R. Hoehe et al. "Sequence Variability and Candidate Gene Analysis in Complex Disease: Association of Mu Opioid Receptor Gene Variation With Substance Dependence," *Human Molecular Genetics* 9:2895–2908.

76. Mary Jeanne Kreek. 2001. "Drug Addictions: Molecular and Cellular Endpoints," *Annals of the New York Academy of Sciences* 937:27–49.

77. Kreek, 2002, p.100.

78. Lee M. Silver. 1997. *Remaking Eden: Cloning and Beyond in a Brave New World*. New York: Avon Books, pp.4–7.

79. Silver, 1997, p.11.

80. Leon Kass. 1988. *Towards a More Natural Science: Biology and Human Affairs*. New York: Free Press, p.25.

81. Daniel Kelves. 1985. *In the Name of Eugenics: Genetics and the Uses of Human Heredity*. Cambridge, MA: Harvard University Press, pp.286–287.

82. Erik Parens and Adrienne Asch. 1999. "The Disability Rights Critique of Prenatal Testing: Reflections and Recommendations," *Hastings Center Report* 29:S1-S22.

83. Norman Daniels. 2000. "Normal Functioning and the Treatment-Enhancement Distinction," *Cambridge Quarterly of Healthcare Ethics* 9:309–322.

84. Masato Senoo et al. 2000. "Adenovirus-Mediated *In Utero* Gene Transfer in Mice and Guinea Pigs: Tissue Distribution of Recombinant Adenovirus Determined by Quantitative TaqMan-Polymerase Chain Reaction Assay," *Molecular Genetics and Metabolism* 69:269–276.

85. LeRoy Walters and Julie Gage Palmer. 1997. *The Ethics of Human Gene Therapy*. New York: Oxford University Press, p.62.

86. Paul Root Wolpe. 2002. "Bioethics, the Genome, and the Jewish Body," *Conservative Judaism* 54: 14–25.

87. Edmund D. Pellegrino. 2001. "The Human Genome Project: The Central Ethical Challenge," *St. Thomas Law Review* 13:815–825.

Chapter 5

1. I. Bernard Cohen. 1985. *Revolution in Science*. Cambridge, MA: The Belknap Press of Harvard University Press, pp.105–145.

2. Charles Darwin. 1859. *The Origin of Species*. New York: Norton (1975 ed.).

3. The Gallup Organization. *Substantial Numbers of Americans Continue to Doubt Evolution as Explanation for Origin of Humans*. Gallup Poll. (March 5, 2001).

4. Svaate Pääbo. 2003. "The Mosaic That Is Our Genome," *Nature* 421:409–412.

5. Marcus W. Feldman, Richard C. Lewontin, and Mary-Claire King. 2003. "A Genetic Melting Pot," *Nature*. 424: 374; L. Luca Cavallisforza, Paolo Menozzi, and Alberto Piazza. 1994. *The History and Geography of Human Genes*. Princeton; Princeton University Press, p.19.

6. Victoria Griffith. 2002. "Wires Cross Over Genes As Information On Ethnic Groups Pours In," *Financial Times (London)* (November 2, 2002) p.1.

7. Comments of J. Craig Venter at the Gene Media Forum, New York, NY, July 20, 2000.

8. Cornell West. 1993. *Race Matters*. Boston: Beacon Press; William Julius Wilson, 1990. *The Truly Disadvantaged: The Inner City, The Underclass, and Public Policy*. Chicago: University of Chicago Press.

9. The International SNP Map Working Group. 2001. "A Map of Human Genome Sequence Variation Contain 1.42 Million Single Nucleotide Polymorphisms," *Nature* 409:928–933. Mary Jeanne Kreek, "Gene Diversity in the Endorphin System," *The Genomic Revolution: Unveiling the Unity of Life*. Michael Yudell and Rob DeSalle, eds. Washington, DC: Joseph Henry Press, 2002. pp.97–98.

10. Joseph L. Graves Jr. 2001. *The Emperor's New Clothes: Biological Theories of Race at the Millennium*. New Brunswick, NJ: Rutgers University Press, p.35.

11. Audrey Smedley. 1993. *Race in North America: Origin and Evolution of a Worldview*. San Francisco, CA: Westview Press, p.165.

12. Smedley, 1993, pp.163–164.

13. Stephen Jay Gould. 1996. *The Mismeasure of Man*. New York: W.W. Norton and Company, pp.69, 410.

14. Thomas Jefferson. 1995 ed. *Notes on the State of Virginia*. Chapel Hill, NC: University of North Carolina Press, pp.138–139.

15. Eugene A. Foster et al. 1998. "Jefferson Fathered Slave's Last Child," *Nature* 396:27–28.

16. George Fredrickson. 1981. *White Supremacy: A Comparative Study in American and South African History*. New York: Oxford University Press, pp.101, 129.

17. Drew Gilpen Faust. 1981. *The Ideology of Slavery: Proslavery Thought in the Antebellum South, 1830–1860*, Baton Rouge, LA: Louisiana State University Press, p.237.

18. Smedley, 1993, p.239.

19. Gould, 1996, p.83

20. Gould, 1996, pp.82–101.

21. Gould, 1996, pp.102–105.

22. Richard Hofstadter. 1992 ed. *Social Darwinism in American Thought*. Boston: Beacon Press.

23. William B. Provine. 1986. "Genetics and Race," *American Zoologist* 26:857–887.

24. Henry Louis Gates, Jr. 1990. "Critical Remarks," *Anatomy of Racism*. David Theo Goldberg, ed., Minneapolis, MN: University of Minnesota Press, p.326; Pierre Van den Berghe. 1978. "Race and Ethnicity: A Sociobiological Perspective," *Ethnic and Racial Studies* 1:404; David Barash. 1979. *The Whisperings Within*. New York: Harper and Row, pp.154, 232.

25. F. James Davis. 1991. *Who Is Black? One Nation's Definition*. University Park, PA: Pennsylvania State University Press, pp.9–11; Calvin Trillin. 1986. "American Chronicles: Black or White," *New Yorker* (April 14, 1986), pp.62–78.

26. Carol C. Mukhopadhyay and Yoland T. Moses. 1997. "Reestablishing "Race" In Anthropological Discourse," *American Anthropologist* 99:517–533.

27. Ashley Montagu. 1997. *Man's Most Dangerous Myth: The Fallacy of Race*. Walnut Creek, CA: AltaMira Press.

28. Ashley Montagu. 1972. *Statement On Race: An Annotated Elaboration and Exposition Of The Four Statements On Race Issued By The United Nations Educational, Scientific, And Cultural Organization*. New York: Oxford University Press, p.10.

29. Guido Barbujani, Arianna Magagni, Eric Minch, and L. Luca Cavalli-Sforza. 1997. "An Apportionment of Human DNA Diversity," *Proceedings of the National Academy of Sciences USA* 94:4516–4519.

30. Neil Risch, Esteban Burchard, Elad Ziv, and Hua Tang. 2002. "Categorizations of Humans in Biomedical Research: Genes, Race, and Disease," *Genome Biology* 2:comment2007.1–comment2007.12.

31. Feldman, Lewontin, and King, 2003, p.374.

32. James F. Wilson et al. 2001. "Population Genetic Structure of Variable Drug Response," *Nature Genetics* 29:265–269.

33. Esteban J. Parra et al. 1998. "Estimating African-American Admixutre Proportions by Use of Population Specific Alleles," *American Journal of Human Genetics* 63:1839–1851; Manfred Kayser et al. 2003. "Y Chromosome STR Haplotypes and the Genetic Structure of U.S. Populations of African, European, and Hispanic Ancestry," *Genome Research* 13:624–634.

34. Mark D. Shriver. 2003. "Skin Pigmentation, Biogeographical Ancestry and Admixture Mapping," *Human Genetics* 112:387–399.

35. A. Chakravarti and R. Chakraborty. 1978. "Elevated Frequency of Tay-Sachs Disease Among Ashkenazic Jews Unlikely By Genetic Drift Alone," *American Journal of Human Genetics* 30:256–261; L.Luca Cavalli-Sforza. 1979. "The Ashkenazi Gene Pool: Interpretations," *Genetic Diseases Among Ashkenazi Jews*. R.M. Goodman and A.G. Motulsky, eds. New York: Raven Press, pp. 93–104.

36. Michael Specter. 1999. "Decoding Iceland: The Next Big Medical Breakthroughs May Result From One Scientist's Battle To Map The Viking Gene Pool," *New Yorker* (January 18, 1999) pp.40–51; Kevin O'Sullivan. 1999. "Unease as Iceland Sells its Entire DNA," *The Irish Times* (January 23, 1999) p.10.

37. E. Arnason. 2003. "Genetic Heterogeneity of Icelanders," *Annals of Human Genetics* 67:5–16; A. Helgason et al. 2003. "A Reassessment of Genetic Diversity in Icelanders: Strong Evidence from Multiple Loci for Relative Homogeneity Caused by Genetic Drift," *Annals of Human Genetics* 67:281–297.

38. Henry T. Greely. 2001. "Human Genome Diversity: What About the Other Genome Project?" *Nature Reviews Genetics* 2:222–227; L.Luca Cavalli-Sforza et al. 1991. "Call For

a World-Wide Survey of Human Genetic Diversity: A Vanishing Opportunity for the Human Genome Project," *Genomics* 11:490–491.

39. Tim Radford. 1993. "Gene Map May Fuel Racism, Scientist Warns," *The Ottawa Citizen* (April 10, 1993), p.D9.

40. Jennifer Couzin. 2002. "New Mapping Project Splits the Community," *Science* 296:1391–1392; David Adam. 2001. "Genetics Group Targets Disease Markers in the Human Sequence," *Nature* 412:105.

41. Pääbo, 2003, pp.409–412.

42. Gillian C.L. Johnson et al. 2001. "Haplotype Tagging For The Identification of Common Disease Genes," *Nature Genetics* 29:233–237; Stacey B. Gabriel. 2002. "The Structure of Haplotype Blocks in the Human Genome," *Science* 296:2225–2229; Lon R. Cardon and Gonçalo R. Abecasis. 2003. "Using Haplotype Blocks to Map Human Complex Trait Loci," *Trends in Genetics.* 19:135–140

43. Couzin, 2002, pp.1391–1392.

44. Morris W. Foster and Richard R. Sharp. 2002. "Race, Ethnicity, and Genomics: Social Classifications as Proxies of Biological Heterogeneity," *Genome Research* 12:844–850.

45. Eliot Marshall. 1998. "DNA Studies Challenge the Meaning of Race," *Science* 282:654–655; Alan R. Temptleton. 2002. "Out of Africa Again and Again," *Nature* 416:45–51.

46. Sharon Schmickle. 2000. "Africa is the Cradle of the Human Race, Scientists Say," *Star Tribune* (July 25, 2000), p.1A; L.B. Jorde et al. 2000. "The Distribution of Human Genetic Diversity: A Comparison of Mitochondrial, Autosomal, and Y-Chromosome Data," *American Journal of Human Genetics* 66:979–988.

47. William B Provine. 1986. "Genetics and Race," *American Zoologist* 26:882.

48. Kelly Owens and Mary-Claire King. 1999. "Genomic Views of Human History," *Science* 286:451–453.

Chapter 6

1. Charles Darwin. 1859. *The Origin of Species.* New York: Norton (1975 ed.).

2. T.A. Brown. 2002. *Genomes 2.* New York: John Wiley and Sons; Daniel Hartl. 1998. *A Primer in Population Genetics.* Sunderland, MA: Sinauer Associates.

3. Jerome Lawrence and Robert Edwin Lee. 1958. *Inherit the Wind.* New York: Dramatists Play Service.

4. Stephen Jay Gould. 1999. *Rocks of Ages: Science and Religion in the Fullness of Life.* New York: Ballantine, p.134.

5. Edward Larson. 1997. *Summer for the Gods: The Scopes Trial and America's Continuing Debate Over Science and Religion.* Cambridge, MA: Harvard University Press, p.83.

6. Larson, 1997, pp.88–92.

7. Larson, 1997, pp.89–91.

8. Larson, 1997, p.27.

9. Larson, 1997, p.94.

10. Paul Conkin, *When All the Gods Trembled: Darwinism, Scopes, and American Intellectuals*, (New York, 1998) pp.79–109.

11. Larson, 1997, p.191.

12. Gould, 1999, p.137.

13. Larson, 1997, pp.187–193.

14. Larson, 1997, pp.7, 134, 177.

15. The Gallup Organization. *Substantial Numbers of Americans Continue to Doubt Evolution as Explanation for Origin of Humans*. Gallup Poll. (March 5, 2001).

16. J.A. Lake. 1985. "Evolving Ribosome Structure: Domains in Archaebacteria, Eubacteria, Eocytes and Eukaryotes," *Annual Review of Biochemistry*. 54:507–530.

17. Tree of Life web project: *http://tolweb.org/tree/phylogeny.html*

18. Carolus Linnaeus. 1735. *Systema Naturae*. Nieuwkoop, The Netherlands:B. de Graaf (1964 ed.).

19. Ewan Birney et al. 2001. "Mining The Draft Human Genome," *Nature* 409:827–828

20. Carl R. Woese et al. 1990. "Towards a Natural System of Organisms: Proposal For the Domains Archaea, Bacteria, and Eucarya," *Proceedings of the National Academy of Sciences USA* 87:4576–4579.

21. National Center for Biotechnology Information, Entrez Genomes: *http://www.ncbi.nlm. nih.gov/PMGifs/Genomes/allorg.html*

22. EuGenes, Genomic Information for Eukaryotic Organisms: *http://eugenes.org:7072/all/ hgsummary.html*

23. Walter J. Gehring and Frank Ruddle. 1998. *Master Control Genes in Development and Evolution: The Homeobox Story*. New Haven, CT: Yale University Press.

24. Takaski Kondo et al. 1997. "Of Fingers, Toes and Penises," *Nature* 390:29.

25. Mark S. Boguski. 2002. "The Mouse That Roared," *Nature* 420:515–516.

26. J.S. Cavanna. 1988. "Molecular and Genetic Mapping of the Mouse MDX Locus," *Genomics* 3:337–341.

27. Robert S. Wilden et al. 2001. "X-Linked Neonatal Diabetes Mellitus, Enteropathy and Endocrinopathy Syndrome is the Human Equivalent of Mouse Scurfy," *Nature Genetics* 27:18–20.

28. The BLAST website can be found at http://*www.ncbi.nlm.nih.gov/BLAST/*

29. Boguksi, 2002, pp.515–6.

30. Marie C. Hogan et al. 2003. "*PKHDL1*, A Homolog of the Autosomal Recessive Polycystic Kidney Disease Gene, Encodes a Receptor With Inducible T Lymphocyte Expression," *Human Molecular Genetics* 12:685–698.

31. David Cyranoski, 2002. "Almost Human..." *Nature* 418:910–912; Asao Fujiyama et al. 2002. "Construction and Analysis of a Human-Chimpanzee Comparative Clone Map," *Science* 295:131–134.

32. Cyranoski, 2002, p.910.

33. Cyranoski, 2002, p.911.

34. Cyranoski, 2002, p.912; Dennis Bray. 2003. "Molecular Prodigality," *Science* 299:1189–1190.

35. Wray et al. 1995. "Genetic Divergence and Geographic Diversitification in *Nautilus*," *Paleobiology* 21:220–228.

36. Environmental News Network. 1998. "Florida Manatees Share Small Gene Pool," *Cnn.com.* (November 18, 1998), *http://www.cnn.com/TECH/science/9811/19/manatee.yoto/*

37. S.A. Nadin-Davis et al. 1996. "The Design of Strain-Specific Polymerase Chain Reactions for Discrimination of the Racoon Rabies Virus Strain from Indigenous Rabies Viruses of Ontario," *Journal of Virology Methods 57:1–14*; Tom Spears. 2003. "DNA Tests Aim to Stop Rabies at Border," *Ottawa Citizen* (January 17, 2003), p.A7.

38. Sarah Graham. 2001 "Protecting St. Vincent Amazon Parrots: DNA 'Featherprinting' May Help Scientists to Thwart the Illegal Trade of a Threatened Species of Parrot," *Scientific American* (September 4, 2001).

39. Rob DeSalle and Vadim Birstein. 1996. "PCR Identification of Black Caviar," *Nature* 381:197.

40. H.C. Rosenbaum et al. 1995. "Paternity Assessments for a Captive Breeding Group of Beluga Whales," Unpublished Paper Presented at The International Association For Aquatic Animal Medicine, 1995.

41. H.C. Rosenbaum et al. 2000. "Worldwide Genetic Differentiation of*Eubalaena*: Questioning the Number of Right Whale Species," *Molecular Ecology* 9:1793–1802.

42. Rosenbaum et al., 2000.

43. Rosenbaum et al., 2000.

44. David W. Oldach et al. 2000. "Heteroduplex Mobility Assay-Guided Sequence Discovery: Elucidation of the Small Subunit (18S) rDNA Sequences of *Pfiesteria piscicida* and Related Dinoflagellates from Complex Algal Culture and Environmental Sample DNA Pools," *Proceedings of the National Academy of Sciences USA* 11:4303–4308.

Chapter 7

1. John C. Sheehan. 1982. *The Enchanted Ring: The Untold Story of Penicillin.* Cambridge, MA: The MIT Press, p.3.

2. Thomas E. Starzl. 2000. "History of Clinical Transplantation," *World Journal of Surgery* 24:759–782.

3. Brad Rodu and Philip Cole. 2001. "The Fifty-Year Decline of Cancer in America," *Journal of Clinical Oncology* 19:239–241.

4. _____, "Achievements in Public Health, 1900–1999: Control of Infectious Diseases," *Morbidity and Mortality Weekly Report*, 48(29):621–629; Gregory L. Armstrong, Laura A. Conn, and Robert W. Pinner, "Trends in Infectious Disease Mortality in the United States During the 20th Century," *JAMA* 281:61–66.

5. Marta Gwinn and Muin J. Khoury, 2002. "Research Priorities for Public Health Sciences in the Postgenomic Era," *Genetics in Medicine* 4:410–411.

6. George Poste. 1998. "Molecular Medicine and Information-Based Targeted Healthcare," *Nature Biotechnology* 16:19–21.

7. _____. "Genome Science Advances to Benefit Third World," *Times of India.* June 14, 2003.

8. World Health Organization, "Genome Research Can Save Millions in Developing World." April 30, 2002. *http://www.who.int/inf/en/pr-2002–34.html*

9. Harold Varmus, 2002. "Getting Ready for Gene-Based Medicine," *New England Journal of Medicine* 347:1526–1527.

10. Kenneth Olden and Samuel Wilson, "Environmental Health and Genomics: Visions and Implications," *Nature Reviews Genetics* 1:149–153.

11. Niall Quinn and Anette Schrag. 1998. "Huntington's Disease and Other Choreas," *Journal of Neurology* 245:709–716.

12. For more on microarray function and use see: Daniel D. Shoemaker and Peter S. Linsley. 2002. "Recent Developments in DNA Microarrays," *Current Opinion in Microbiology* 5:334–337; A.P. Levene et al. 2003. "The Uses of Genetic Microarray Analysis to Classify and Predict Prognosis in Haematological Malignancies," *Clinical and Laboratory Haematology* 25:209–220.

13. Tracy Webb. 2003. "Microarray Studies Challenge Theories of Metastasis," *Journal of the National Cancer Institute* 95: 350–351; Marc J. van de Vijver and Yudong D. He. 2003. "A Gene-Expression Signature as a Predictor of Survival in Breast Cancer," *New England Journal of Medicine* 347:1999–2009.

14. Ash A. Alizadeh, Michael B Eisen, et al. 2000. "Distinct Types of Diffuse Large B-Cell Lymphoma Identified by Gene Expression Profiling," *Nature* 403:503–511; Andreas Rosenwald, George Wright, Wing C. Chan. 2002. "The Use of Molecular Profiling to Predict Survival After Chemotherapy for Large B-Cell Lympoma," *New England Journal of Medicine* 346:1937–1947.

15. Arnold J. Levine. 2001. "The Origins of Cancer and the Human Genome," In *The Genomic Revolution: Unveiling the Unity of Life*. Michael Yudell and Rob DeSalle, eds. Washington, DC: Joseph Henry Press, p.94.

16. Thomas S. May. 2003. "Gleevec: Tailoring to Fit," *Drug Discovery Today* 8:188–189.

17. Michael E. O'Dwyer, Michael J. Mauro, and Brian J. Drucker. 2002. "Recent Advances in the Treatment of Chronic Myelogenous Leukemia," *Annual Review of Medicine* 53:369–381.

18. Hagop Kantarjian. 2002. "Hematologic and Cytogenic Responses to Imatinib Mesylate in Chronic Myelogenous Leukemia," *New England Journal of Medicine* 346:645–652.

19. Janet Fricker. 2001. "Genome Sequences Reveal Key Genetic Element in PD," *Drug Discovery Today* 6:277–278.

20. Sui Huang. 2002. "Rational Drug Discovery: What Can We Learn From Regulatory Networks?" *Drug Discovery Today* 7:s163–s169.

21. Mike Tyers and Matthias Mann. 2003. "From Genomics to Proteomics," *Nature* 422:193–197.

22. Alison Abbot. 2001. "Workshop Prepares Ground for Human Proteome Project," *Nature* 409:747.

23. Genetests, 2003. *http://www.genetests.org*

24. The New York State Task Force on Life and the Law. 2000. *Genetic Testing and Screening in the Age of Genomic Medicine.* p.30.

25. The New York State Task Force on Life and the Law, 2000, pp.141–144.

26. For more information on Canavan disease see: *http://www.geneclinics.org/profiles/canavan/* and *http://www.canavanfoundation.org/*

27. Michael M. Burgess. 2001. "Beyond Consent: Ethical and Social Issues in Genetic Testing," *Nature Reviews Genetics* 2:147–151.

28. Karen Greendale and Reed E. Pyeritz. 2001. "Empowering Primary Care Health Professionals in Medical Genetics," *American Journal of Medical Genetics* 106:223–232; Wylie Burke and Jon Emery. 2002. "Genetics Education for Primary Care Providers," *Nature Reviews Genetics* 3:561–566; Michael L. Begleiter. 2002. "Training for Genetic Counsellors," *Nature Reviews Genetics* 3:557–561.

29. The New York State Task Force on Life and the Law, 2000, p.111.

30. Katherine N. Nathanson, Richard Wooster, and Barbara L. Weber. 2001. "Breast Cancer Genetics: What We Know and What We Need," *Nature Medicine* 7:552–556.

31. Wadman, 1998, p.851. Ashkenazi Jewish women, for example, have a higher frequency of the BRCA-1 and BRCA-2 genes, which have been linked to an increased incidence of breast and ovarian cancers.

32. Antonio Regalado. 2002. "Want a Map of Your DNA?" *Chicago Sun Times* (October 6, 2002) p.36.

33. Francis S. Collins, Eric D. Green, Alan E. Guttmacher, and Mark S. Guyer on Behalf of the US National Human Genome Research Institute. 2003. "A Vision for the Future of Genomics Research: A Blueprint for the Genomic Era." *Nature* 422:835–847.

34. Barbara Bowles Biesecker and Teresa M. Marteau. 1999. "The Future of Genetic Counseling: An International Perspective," *Nature Genetics* 22:133–137.

35. Lori B. Andrews. 2001. *Future Perfect: Confronting Decisions About Genetics.* New York: Columbia University Press, pp.6–7.

36. Kathryn A. Phillips et al. 2001. "Potential Role of Pharmacogenomics in Reducing Adverse Drug Reactions," *JAMA* 286:2270–2279.

37. Allen D. Roses. 2000. "Pharmacogenomics and the Practice of Medicine," *Nature* 405:857–865.

38. Phillips, et al., 2001, pp.2270–2279.

39. Richard Weinshilboum. 2003. Genomic Medicine: Inheritance and Drug Response," *New England Journal of Medicine* 348:529–537.

40. Mary V. Reiling and Thierry Dervieux. 2001. "Pharmacogenetics and Cancer Therapy," *Nature Reviews Cancer* 1:99–108; R Nagasubramanian et al. 2003. "Pharmacogenetics in Cancer Treatment," *Annual Reviews Medicine* 54:437–452.

41. SNP Consortium Website: *http://snp.cshl.org/*

42. Eshan Masood. 1999. ". . . as Consortium Plans Free SNP Map of Human Genome," *Nature* 398:545–546.

43. James F. Wilson, et al. 2001. "Population Genetic Structure of Variable Drug Response." *Nature Genetics* 29:265–269.

44. Guido Barbujani, Arianna Magagni, Eric Minch, and L. Luca Cavali-Sforza. 1997. "An Apportionment of Human DNA Diversity," *Proceedings of the National Academy of Sciences USA* 94:4516–4519.

45. Roses, 2000, pp.857–865.

46. Buchanan et al. 2002. "Pharmacogenetics: Ethical Issues and Policy Options," *Kennedy Institute of Ethics Journal* 12:1–15.

47. Arthur Caplan. 1997. "The Human Genome Project and the Future of Health Care," *Journal of Legal Medicine* 18:547–551.

48. Kenneth Olden and Samuel Wilson. 2000. "Environmental Health and Genomics: Visions and Implications," *Nature Reviews Genetics* 1:149–153.

49. For example, two of the early and most influential studies of the link between smoking and lung cancer were E.L. Wynder and E. Graham. 1950. "Tobacco Smoking as a Possible Etiologic Factor in Bronchiogenic Carcinoma: A Study of 684 Proven Cases," *JAMA* 143:329–336; R. Doll and A. Bradford Hill. 1956. "Lung Cancer and Other Causes of Death in Relation to Smoking," *British Medical Journal* 2:1071–1081.

50. An initial study of middle-aged smokers, for example, suggests that those with a particular genetic variant of the E4 allele have a threefold risk of a coronary heart disease event compared to those without the allele. Steve E. Humphries, Philippa J. Talmud, Emma Hawet et al. 2001. "Apolipoprotein E4 and Coronary Heart Disease in Middle-Aged Men Who Smoke: A Prospective Study," *The Lancet* 358:115–111.

51. Kenneth Olden and Samuel Wilson, 2000, pp.149–153; Kenneth Olden, Janet Guthrie, and Sheila Newton. 2001. "A Bold New Direction for Environmental Health Research," *American Journal of Public Health* 91:1964–1967.

52. Emile F. Nuwaysir, Michael Bittner, Jeffrey Trent et al. 1999. "Microarrays and Toxicology: The Advent of Toxicogenomics," *Molecular Carcinogenesis* 24:153–159.

53. National Institute of Environmental Health Sciences Environmental Genome Project website, "About EGP": *http://www.niehs.nih.gov/envgenom/egp3.htm.*

54. Jocelyn Kaiser. 2003. "Tying Genetics to the Risk of Environmental Diseases," *Science* 300:563.

55. Report on Carcinogens, Tenth Edition. 2002. U.S. Department of Health and Human Services, Public Health Service, National Toxicology Program. *http://ehp.niehs.nih.gov/roc/toc10.html*

56. Kenneth Olden et al. 2001. pp.1964–1967.

57. Richard A Lovett. 2000. "Toxicologists Brace for Genomics Revolution," *Science* 289:536–537.

58. Jocelyn Kaiser. 2001. "Hunting for Collaborators of Killer Toxins," *Science* 291:1207.

59. Merrill C. Miller, III, Harvey W. Mohrenwiser, and Douglas A. Bell, 2001. "Genetic Variability in Susceptibility and Response to Toxicants," *Toxicology Letters* 120:269–280.

60. Giovanna Lombardi, Conrad Germain et al. 2001. "HLA-DP Allele Specific T Cell Responses to Beryllium Account for DP-Associated Susceptibility to Chronic Beryllium Disease," *Journal of Immunology* 166:3549–3555; Zaolin Wang et al. 1999. "Differential Susceptibilities to Chronic Beryllium Disease Contributed by Different GLU69 HLA-DPB1 and –DPA1 Alleles," *Journal of Immunology* 163:1647–1653.

61. U.S. Department of Energy. 2000. "New Tests Will Help Predict, Diagnose Chronic Beryllium Disease," *Dateline: Los Alamos* (April 2000).

62. Stephanie Armour. 1999. "Many Workers Fear Genetic Discrimination by Employers," *USA Today* (May 5, 1999) p.2B.

63. David C. Christiani, Richard R. Sharp, Gwen W. Collman, and William A. Suk. 2001. "Applying Genomic Technologies in Environmental Health Research: Challenges and Opportunities," *Journal of Occupational and Environmental Medicine* 43:526–533.

64. Alexander Pfeifer and Inder M. Verma, 2001. "Gene Therapy: Promises and Problems," *Annual Review of Genomics and Human Genetics* 2:177–211.

65. Leroy Walters and Julie Gage Palmer. 1997. *The Ethics of Human Gene Therapy*. New York: Oxford University Press, pp.166–167.

66. Peter R. Vale, Douglas Losordo et al., 2000. "Left Ventricular Electromechanical Mapping to Assess Efficacy of phVEGF Gene Transfer for Therapeutic Angiogenesis in Chronic Myocardial Ischemia," *Circulation* 102:965–974; Dan Ferber. 2001. "Gene Therapy: Safer and Virus-Free?" *Science* 294:1638–1642.

67. Clare E. Thomas, Anja Ehrhardt, and Mark A. Kay. 2003. "Progress and Problems with the Use of Viral Vectors for Gene Therapy," *Nature Reviews Genetics* 4:346–358.

68. Alessandro Aiuti, Shimon Slavin, Memet Aker et al. 2002. "Correction of ADA-SCID by Stem Cell Gene Therapy Combined with Nonmyeloablative Conditioning," *Science* 296:2410–2413.

69. Sherly Gay Stolberg. 1999. "The Biotech Death of Jesse Gelsinger," *The New York Times Magazine* (November 28, 1999) p.137.

70. Xiaolin We, Yan Li, Bruce Crise, and Shawn M. Burgess. 2003. "Transcription Start Regions in the Human Genome Are Favored Targets for MLV Integration," *Science* 300:1749–1751.

71. Karol Sikora. 1999. "Introduction," *Gene Therapy: Principles and Applications*, Ed. Thomas Blankenstein. Boston: Birkhäuser Verlag, pp.1–10.

72. Federica Sangiuolo et al. 2002. "*In Vitro* Correction of Cystic Fibrosis Epithelial Cell Lines by Small Fragment Homologous Replacement (SFHR) Technique," *BMC Medical Genetics* 3:8.

73. Eliot Marshall. 2001. "Gene Gemisch Cures Sickle Cell in Mice," *Science* 294:2268; Robert Pawliuk et al. 2001. "Correction of Sickle Cell Disease in Transgenic Mouse Models by Gene Theraphy," *Science* 294:2268.

74. Robert Pawliuk, Karen A Westerman et al. 2001. "Correction of Sickle Cell Disease in Transgenic Mouse Models by Gene Therapy," *Science* 294:2368–2371

75. Marina Cavazzana-Calvo, Salima Hacein-Bey, Geneviève de Saint Basile et al. 2000. "Gene Therapy of Human Severe Combined Immunodeficiency (SCID)-X1 Disease," *Science* 288:669–672.

76. Brendan A Maher. 2002. "Gene Therapy Trials Hit Obstacle," *The Scientist* 16:26; Jeffrey L. Fox. 2003. "FDA Panel Recommends Easing Gene Therapy Trial Limits," *Nature Biotechnology* 21:344–345; Nathan R.Wall and Yang Shi. 2003. "Small RNA: Can RNA Interference be Exploited for Therapy?" *The Lancet* 362:1401–1403; Beverly L. Davidson and Brain: Silencing of Disease Genes with RNA Interference," *The Lancet Neurology*. 3:145-149; Erika Check. 2004. "Gene Therapists Hopeful as Trials Resume With Childhood Disease," *Nature* 429:587.

Chapter 8

1. The Melman Group, Inc. 2003. "Public Sentiments About Genetically Modified Food." *http://pewagbiotech.org/research/2003update/*

2. Pew Initiative on Food and Biotechnology. 2004. "Feeding The World: A Look at Biotechnology and World Hunger" March 2004. *http://pewagbiotech.org*; Mike Toner. 2002. "Eating Altered Genes," *Atlanta Journal and Constitution* (May 19, 2002) p.1A.

3. Henry A. Wallace and William L. Brown. 1988. *Corn and Its Early Fathers*. Ames: Iowa State University Press, p.31.

4. Michael Specter. 2000. "The Pharmageddon Riddle," *New Yorker* (April 10, 2000) p.58.

5. Specter, 2000, p.69.

6. Paul Lurquin. 2001. *The Green Phoenix: A History of Genetically Modified Plants*. New York: Columbia University Press, pp.56–85.

7. Lurquin, 2001, pp.98–101.

8. Lurquin, 2001, pp.96–97

9. Pew Initiative on Food and Biotechnology, 2003.

10. Charles C. Mann. 1999. "Crop Scientists Seek a New Revolution," *Science* 283:310–314; David Tilman et al. 2001. "Forecasting Agriculturally Driven Global Environmental Change," *Science* 292:281–284; Jikun Huang et al. 2002. "Enhancing Crops to Feed the Poor," *Nature* 418:687–684; Pew Initiative on Food and Biotechnology, 2004.

11. Sharon Schmickle. 1999. "Bug Warfare," *Star Tribune (Minneapolis)* (March 17, 1999) p.8A.

12. Paul Lurquin. 2002. *High Tech Harvest: Understanding Genetically Modified Food Plants*. Boulder, CO: Westview Press, pp.104–105.

13. Paul Christou. 1996. *Particle Bombardment For Genetic Engineering of Plants*. Austin, TX: Academic Press, p.66; M.G. Koziel et al. 1993. Field Performance of Elite Transqenic

Maize Plants Expressing an Insecticidal Protein Derived From *Bacillus thuringiensis,*" *Biotechnology* 11:194–200.

14. John E. Losey et al. 1999. "Transgenic Pollen Harms Monarch Larvae," *Nature.* 399: 214.

15. Richard L. Hellmich et al. 2001. "Monarch Larvae Sensitivity to *Bacillus thuringiensis* Purified Proteins and Pollen," *Proceedings of the National Academy of Sciences USA* 98:11925–11930; Sharon Schmickle. 2002. "Biotech Corn Hazard Low, Early Tests Show," *Star Tribune (Minneapolis)* (November 18, 2002) p.1A.

16. Juan Pablo, Ricard Martínez-Soriano, and Diana Sara Leal-Klevezas. 2000. "Transgenic Maize in Mexico: No Need for Concern," *Science* 287:1399; Rex Dalton. 2001. "Transgenic Corn Found Growing in Mexico," *Nature* 413:337.

17. Lurquin, 2002, p.99; Adrian Slater, Nigel W. Scottt, and Mark R. Fowler et al. 2003. *Plant Biotechnology: The Genetic Manipulation of Plants.* New York: Oxford University Press, pp.112–119.

18. Lurquin, 2002, pp.99–100.

19. Slater et al., 2003, pp.126–129.

20. Slater et al., 2003, pp.247–251.

21. Michael Pollan. 2001. "The Great Yellow Hype," *New York Times Magazine* (March 4, 2001) p.15; Jon Christensen. 2000. "Scientists at Work," *New York Times* (November 21, 2000) p.1F; John Mason. 2003. "Regulators Hinder Modified Rice," *Financial Times (London)* (March 21, 2003) p.11.

22. Biing-Hwan Lin et al. 2001. "Fast Food Growth Boosts Frozen Potato Consumption," *FoodReview* 24:38–46; Michael Durham. 1996. "Look What's Coming to Dinner...Scrambled Gene Cuisine," *The Observer* (October 6, 1996) p.14; Committee on Biobased Industrial Products. 2000. *Biobased Industrial Products: Priorities for Research and Commercialization.* Washington, DC: National Academies Press, pp.48–49.

23. Damian Carrington. 1999. "Fluorescent GM Potatoes Say 'Water Me,'" *BBC News Online* (September 14, 1999): *http://news.bbc.co.uk/2/hi/sci/tech/specials/sheffield_99/446837.stm*; Anjana Ahuja. 1995. "Why We Need Crops That Glow in the Dark," *The Times (London).* (November 27, 1995) p.16.

24. Mike Toner. 2003. "Scientists Create Decaf Coffee Plants," *Atlanta Journal and Constitution* (June 19, 2003) p.13A.

25. John Ohlrogge and John Browse. 1998. "Manipulation of Seed Oil Production," *Transgenic Plant Research.* Keith Lindsey ed. Amsterdam: Harwood Academic Publishers, pp.151–174; Lurquin, 2002, p.126.

26. Glynis Giddings et al. 2000. "Transgenic Plants as Factories for Biopharmaceuticals," *Nature Biotechnology* 18:1151–1155.

27. Giddings, 2000, pp.1154–1155.

28. Carol Kaesuk Yoon. 2000. "Redesigning Nature," *New York Times* (May 1, 2000) pA1.

29. Sharon Schmickle and Rob Hotakainen. 2001. "The Big One That's On Its Way?" *Star Tribune (Minneapolis)* (May 14, 2001) p.1A; Alwyn Scott. 2003. "Will Europe's Food Fears Turn Tastes to Wild Fish?" *The Seattle Times* (May 25, 2003) p.E1.

30. Warren E. Leary. 2002. "Panel Urges Caution in Producing Gene-Altered Animals," *New York Times* (August 21, 2002) p.A12; Sharon Schmickle. 2001. "Cloning Debate Moves From Lab to the Barnyard," *Star Tribune (Minneapolis).* (December 2, 2001) p.1A.

31. David Humpherys et al. 2001. "Epigenetic Instability in ES Cells and Cloned Mice," *Science* 293:95–97; Ian Wilmut et al. 2002. "Somatic Cell Nuclear Transfer," *Nature* 419:583–586.

32. Chris Fusco. 2003. "Getting Families to Donate Organs Takes Kind Touch," *Chicago Sun-Times* (August 28, 2003) p.20.

33. Marialuisa Lavitrano. 2002. "Efficient Production by Sperm-Mediated Gene Transfer of Human Decay Accelerating Factor (Hdaf) Transgenic Pigs for Xenotransplantation," *Proceedings of the National Academy of Sciences USA* 9:14230–14235; Naomi Aoki. 2003. "Organ Transplants Get Lift In Cloned Piglet," *Boston Globe* (January 12, 2003) p.E1.

34. Pat Eaton-Robb. 1999. "Altered Pig Parts Offer Humans Hope," *The Ottawa Citizen* (February 23, 1999) p.A9.

35. Kendall Powell. 2003. "Barnyard Biotech–Lame Duck or Golden Goose," *Nature Biotechnology* 21:965–967;

36. Thomas Maeder. 1998. "Tobacco Can Be Good For You," *New York Times* (August 28, 1998) p. 37; _____. 1998. "Scientists Develop Vaccine Against Tooth Decay," CNN Interactive (April 29, 1998): *http://www.cnn.com/HEALTH/9804/29/tooth.decay/*

37. Rebecca Eisenberg. 2002. "How Can You Patent Genes?" *Who Owns Life?* David Magnus et al., eds. New York: Prometheus Books, pp.117–134.

38. Kendall Powell. 2002. "Genes Improve Green Cleaning," *Nature Science Update.* (October 8, 2002); O.P. Dhankher et al. 2002. "Engineering Tolerance and Hyperaccululation of Arsenic in Plants by Combining Arsenate Reductase and γ-Glutamylcysteine Synthetase Expression," *Nature Biotechnology* 20:1140–1145.

39. J. Craig Venter. 2002. "Whole-Genome Shotgun Sequencing," *The Genomic Revolution: Unveiling the Unity of* Life. Washington, DC: Joseph Henry Press, p.51.

40. Eleanor Lawrence. 2002. "Microbial Mercury Mop," *Nature Science Update* (October 8, 2002).

41. H. Brim. 2000. "Engineering *Deinococcus Radiodurans* for Metal Remediation in Radioactive Mixed Waste Environments," *Nature Biotechnology* 18:85–90.

42. _____. 1998. "Illuminating Landmines," *Energy Science News* (January-February 1999) *http://www.pnl.gov/energyscience/*

43. A.R. Watkinson et al. 2000. "Predictions of Biodiversity Response to Genetically Modified Herbicide-Tolerant Crops," *Science* 289:1554–1557.

44. Philip J. Dale et al. 2002. "Potential for the Environmental Impact of Transgenic Crops," *Nature Biotechnology* 20:567–574; Pew Initiative on Food and Biotechnology. 2003. *Have Transgenes, Will Travel: Issues Raised by Gene Flow from Genetically Engineered Crops.* *http://www.pewagbiotech.org*

45. L.L. Wolfenbarger and P.R. Phifer. 2000. "The Ecological Risks and Benefits of Genetically Engineered Plants," *Science* 290:2088–2093.

46. Peter A. Singer and Abdallah S. Daar. 2001. "Harnessing Genomics and Biotechnology to Improve Global Health Equity," *Science* 294:87–89.

47. David Usborne. 1999. "The Monster Within," *The Independent (London)* (April 21, 1999) p.3.

48. Spector, 2000, pp.62–63.

49. Mary Dejevsky and Oliver Tickell. 1999, "Monsanto to Face its Critics as GM Markets Shrink," *The Independent (London).* (October 6, 1999) p.16.

50. Robert F. Service. 1998. "Seed-Sterilizing 'Terminator Technology' Sows Discord," *Science* 282:850–851.

51. Service, 2000, p.850. Monsanto was in the process of acquiring the terminator technology through the purchase of Delta and Pine Land, a small cottonseed company that had patented what it called a "technology protection system"—which came to be known derisively as "terminator technology."

52. U.S. Food and Drug Administration, Center for Food Safety and Applied Nutrition. 1994. "Biotechnology of Food," *FDA Backgrounder* (May 18, 1994) *http://vm.cfsan.fda.gov/~lrd/biopolicy.html.*

53. J.H. Maryanski. 1995. "FDA's Policy for Foods Developed By Biotechnology," *Genetically Modified Foods: Safety Issues.* Engel, Takeoka, and Teranishi eds. American Chemical Society, Symposium Series No. 605. *http://vm.cfsan.fda.gov/~lrd/biotechn.html.*

54. Pew Initiative on Food and Biotechnology. 2001. "Guide to U.S. Regulation of Agricultural Biotechnology Products." *http://pewagbiotech.org/resources/issuebriefs/1-regguide.pdf;* Griffe Witte. 2004. "Rules on Biotech Crops to be Revised," *Washington Post.* (January 23, 2004) p.A2; Pew Initiative on Food and Biotechnology 2004. "Issues in the Regulation of Genetically Engineered Plants and Animals," (April 2004). *http://pewagbiotech.org*

55. Andrew Pollack. 2000. "Kraft Recalls Taco Shells With Bioengineered Corn," *New York Times* (September 23, 2000).

56. Alan McHughen. 2000. *Pandora's Picnic Basket: The Potential and Hazards of Genetically Modified Foods.* New York: Oxford University Press, pp.119–121.

57. Jim Ritter. 2000. "Genetic Products Losing Favor With U.S. Food Firms," *Chicago Sun-Times.* (February 29, 2000) p.4.

58. Andrew Pollack. 2000. "Labeling Genetically Altered Food is a Thorny Issue," *New York Times.* (September 26, 2000), p.A1.

59. Marc Kaufman. 2001. "Biotech Corn Found in a Variety of Foods: FDA Testing for Possible Allergic Reactions," *Washington Post* (April 24, 2001) p.A03; James Cox. 2000. "Starlink Fiasco Wreaks Havoc in the Heartland," *USA Today* (October 27, 2000) p.1B.

60. Ambrose Evans-Pritchard. 2003. "EU Lifts Five-Year Ban on GM Food, But Shoppers Will Have Choice," *Daily Telegraph (London)* (July 2, 2003) p.4.

61. Mike Toner. 2002. "Eating Altered Genes: Designer Crops Already on Grocery Shelves," *Atlanta Journal-Constitution* (May 19, 2002) p.1A.

Chapter 9

1. Gerald Holton. 1996. *Einstein, History, and Other Passions: The Rebellion Against Science at the End of the Twentieth Century*. Cambridge, MA: Harvard University Press, p.125.

2. Fred Jerome. 2002. *The Einstein File: J. Edgar Hoover's Secret War Against the World's Most Famous Scientist*. New York: St. Martin's Press, pp.33–34.

3. Denis Brian. 1996. *Einstein: A Life*. New York: John Wiley & Sons, Inc., p.420; Alice Caprice. 1996. *The Quotable Einstein*. Princeton: Princeton University Press, p.128.

4. Troy Duster. 2002. "Social Side Effects of the New Human Molecular Genetic Diagnostics," *The Genomic Revolution: Unveiling the Unity of Life*. Michael Yudell and Rob DeSalle, eds. Washington, DC: Joseph Henry Press, pp.184–192.

5. Robert Williamson and Rony Duncan. 2002. "DNA Fingerprinting For All," *Nature* 418:585–586.

6. Andrew Pollack. 2003. "A Revolution at 50: How the Arms of the Helixes Are Poised to Serve," *New York Times*. (February 25, 2003) p.5.

7. Identifications are being carried out by the International Commission On Missing Persons: *http://www.ic-mp.org/icmp/home.php*

8. Williamson and Duncan, 2002, p.585.

9. Rebecca Fowler. 2003. "Coded Revelations: DNA The Second Revolution," *The Observer* (April 27, 2003) p.50.

10. Lise Olsen and Roma Khanna. 2003. "DNA Lab Analysts Unqualified," *Houston Chronicle* (September 7, 2003) p.1.

11. Steve McVicker. 2003. "Interim Chief Aims to Clean Up Crime Lab," *Houston Chronicle* (September 23, 2003) p.1.

12. Olsen and Khanna, 2003, p.1.

13. The Innocence Project Website: *http://www.innocenceproject.org*.

14. Philip Reilly. 2001. "Legal and Public Policy Issues in DNA Forensics," *Nature Reviews Genetics* 2:313–317.

15. Kelly Fox and Donna Lyons. 2002. "Fighting Crime With DNA," *National Conference of State Legislatures LegisBrief*. (October 2002). *http://www.ncsl.org*

16. Michelle Hibbert. 1999. "DNA Databanks: Law Enforcement's Greatest Surveillance Tool?" *Wake Forest Law Review* (Fall 1999).

17. Reilly, 2001, pp.313–317.

18. Tom Sheridan. 1997. "Theologians Voice Concern On Cloning," *Chicago Sun-Times*. (March 1, 1997) p.15.

19. Sheryl Gay Stolberg. 2003. "House Votes to Ban All Human Cloning," *New York Times*. (February 28, 2003) p.22.

20. National Conference of State Legislatures website: *http://www.ncsl.org/programs/health/genetics/rt-shcl.htm*

21. J.W. McDonald et al. 2004 "Repair of the Injured Spinal Cord and the Potential of Embryonic Stem Cell Transplantation," *Journal of Neurotrauma* 21:383–393: Linda G. Griffith and Gail Naughton. 2002. "Tissue Engineering Current Challanges and Expanding Opportunities," *Science* 295:1009–1014; Muhammad Al-Hajj et al. 2004. "Therapeutic Implications of Cancer Stem Cells," *Current Opinion in Genetics and Development* 14:43–47; Donald Orlis et al. 2001."Bone Marrow Cells Regenerate Infacted Myocardium," *Nature.* 410:701–705.

22. Søoren Holm. 2002. "Going to the Roots of the Stem Cell Controversy," *Bioethics* 16:493–507; George Annas, Arthur Caplan, and Sheman Elias. 1999. "Stem Cell Politics, Ethics, and Medical Progress," *Nature Medicine* 5:1339–1341.

23. Sheryl G. Stolberg. 2001. "The President's Decision: A Question of Research," *New York Times.* (August 10, 2001) p. A17; Katherine Q. Seelye. 2001. "The Presidents Decision: The Overview: Bush Gives the Backing for Limited Research on Existing Stem Cells," *New York Times.* (August 10, 2001) p. Al; _____. 2003. "Research Makes Adult Stem Cells Multiply," *Cancer Weekly.* (December 16, 2003) p. 47.

24. Stephen S. Hall. 2002. "President's Bioethics Council Delivers," *Science* 297:322–324; Presidents Council on Bioethics. 2002. *Human Cloning and Human Dignity: An Ethical Inquiry.* *http://www.bioethics.gov/reports/cloningreport/index.html.*

25. Bert Vogelstein, Bruce Alberts, and Kenneth Shine. 2002. "Please Don't Call it Cloning," *Science* 295:1237.

26. Rick Weiss. 2003. "Genome Project Completed," *Washington Post* (April 14, 2003) p.6.

PHOTO CREDITS

COVER IMAGE

Denis Finnin, American Museum of Natural History

INTRODUCTION: WELCOME TO THE GENOME

AP Photo/Rick Bowmer, xvii

Jackie Beckett, American Museum of Natural History, xviii

American Museum of Natural History, xxii

Denis Finnin, American Museum of Natural History, xxiii

1: FROM MENDEL TO MOLECULES

Photo Researchers, 3

Photo Researchers, 5

ID Twins, Stone Trujillo-Paumier, Getty Images; Fraternal Twins, Barbara Penoyar, Getty Images, 6

American Museum of Natural History, 9

Principles of Genetics, 3rd Edition, D. Peter Snustad and Michael J. Simmons, © 2003 John Wiley and Sons, Inc. This material is used by permission of John Wiley & Sons, Inc., 11

Visuals Unlimited, 14

Principles of Genetics, 3rd Edition, D. Peter Snustad and Michael J. Simmons, © 2003 John Wiley and Sons, Inc. This material is used by permission of John Wiley & Sons, Inc., 16

DNA Learning Center, Cold Spring Harbor Laboratory, 18

Cold Spring Harbor Laboratory, 22

Cold Spring Harbor Laboratory, 23

Normal Collection for the History of Molecular Biology, 24

Principles of Genetics, 3rd Edition, D. Peter Snustad and Michael J. Simmons, © 2003 John Wiley and Sons, Inc. This material is used by permission of John Wiley & Sons, Inc., 25

2: MAKING DNA PHOTOS

Principles of Genetics, 3rd Edition, D. Peter Snustad and Michael J. Simmons, © 2003 John Wiley and Sons, Inc. This material is used by permission of John Wiley & Sons, Inc., 29

Visuals Unlimited, 30

AP Photo, 31

Visuals Unlimited, 34

Exhibitions Department, American Museum of Natural Hisotory, 35

Exhibitions Department, American Museum of Natural Hisotory, 38

3: HOW GENES WORK

Principles of Genetics, 3rd Edition, D. Peter Snustad and Michael J. Simmons, © 2003 John Wiley and Sons, Inc. This material is used by permission of John Wiley & Sons, Inc., 44

Item from the Neitz Test of Color Vision copyright © 2000 by Western Psychological Services. Reprinted by permission of the publisher, Western Psychological Services, 12031 Wilshire Boulevard, Los Angeles, California, 90025, U.S.A., www.wpspublish.com. Not to be reprinted in whole or in part for any additional purpose without the expressed, written permission of the publisher. All rights reserved, 45

Applied Biosystems, 51

American Museum of Natural History, 52

Zebrafish: Steven Zimmerman and Deborah Yelon, Skirball Institute of Biomolecular Medicine, New York University School of Medicine; C. elegans: David H.A. Fitch; E. coli; American Museum of Natural History, 54

Handbook of Comparative Genetics, Cecilia Saccone and Graziano Pesole, 2003 © Wiley-Liss. This material is used by permission of Wiley-Liss, Inc., a subsidiary of John Wiley and Sons, Inc., 56

4: KEEPING THE GENOME SAFE

Gamma Press, 62

National Library of Medicine / Public Domain, 68

Mary Gelsinger, 71

Mike Thompson / Detroit Free Press, 78

5: 99.9% PHOTOS
Picture Quest, 82

Craig Chesek, American Museum of Natural History, 87

6: THE TREE OF LIFE PHOTOS
The Dallas Morning News, 96

Bettman Images, 98

F. Rudolf Turner, Indiana University, turner@indiana.edu, 103

Lawrence Berkeley National Laboratory, 105

A. Plumptre / Wildlife Conservation Society, 107

Photo by Shannon Crownover, courtesy of Caviar Emptor, 2003, 109

Wildlife Conservation Society, 110

Copyright © Brandon Cole / www.brandoncole.com, 111

7: THE WORLD TO COME: MEDICINE PHOTOS
AP Photo, 120

American Museum of Natural History, 122

Principles of Genetics, 3rd Edition, D. Peter Snustad and Michael J. Simmons, © 2003
   John Wiley and Sons, Inc. This material is used by permission of John Wiley & Sons,
   Inc., 137

8: THE WORLD TO COME: AGRICULTURE PHOTOS
Roderick Mickens, American Museum of Natural History, 142

Principles of Genetics, 3rd Edition, D. Peter Snustad and Michael J. Simmons, © 2003
   John Wiley and Sons, Inc. This material is used by permission of John Wiley & Sons,
   Inc., 143

Photo Researchers, 145

© 2003 IRRI Photo Bank, 147

Aqua Bounty Canada, 149

Roslin Institute, 150

Visuals Unlimited, 152

9: CONCLUSION: IN A GENOMIC WORLD
Edvotek / www.edvotek.com, 163

INDEX

Welcome to the Genome, by Rob DeSalle and Michael Yudell.
ISBN: 0-471-45331-5 Copyright © 2005 Rob DeSalle and Michael Yudell.